MEMOIRS ON WORLD RELIGIONS
Philosophy, Ethics and Comparison

Studies In World Religions

MEMOIRS ON WORLD RELIGIONS
Philosophy, Ethics and Comparison

Studies In World Religions

Babalola Sunday Funmilola
Ekpaogba Sunny Manyo

ELIVA PRESS

ISBN: 978-99949-8-157-1

Cover Design: Eliva Press
Cover Image: Ingimage
Email: info@elivapress.com
Website: www.elivapress.com

This Eliva Press imprint is published by the registered company Eliva Press Global Ltd. part of Eliva Press S.R.L. Publishing Group

The registered company address is: Pope Hennessy Street Level 2, Hennessy Tower Port Louis, Mauritius
Eliva Press S.R.L. Publishing Group address is: Bulevardul Moscova 21, Chisinau, Moldova, Europe

MEMOIRS ON WORLD RELIGIONS

Philosophy, Ethics and Comparison

Sunday Funmilola Babalola

And

Ekpaogba Sunny Manyo

Department of Philosophy and Religious Studies
Joseph Ayo Babalola University,
Ikeji-Arakeji,
Nigeria

Preface

Memoirs on World Religions is written with emphasis on the Philosophy and Ethics of the selected world religions in the atmosphere of Comparison. Many books have surfaced on world religions but not much attention has been paid to the philosophy and ethics of such religions in the atmosphere of comparison. This is the gap that this author intends to fill. In this work, scholarly attempt is made to bring to the fore, the philosophy and ethics of each religion. Beyond that, one religion is compared to another to see where they are similar and where they diverge.

Acknowledgements

I thank God, the owner of heaven and earth for his enablement and grace to write this book. I also appreciate Deborah (my wife), John and David, my children for their support all the way. My appreciation also goes to Dr Sunny Manyo Ekpaogba for his very relevance contribution to this study.

Dedication.

I dedicate this study to God, the owner of heaven and the earth and to all adherents of various religious traditions who accommodate others regardless of their religious faith.

Table of Contents

Chapter One
Introduction

Memoirs on World religions is designed as a teaching note or manual on the theme of study – World religions. The work takes special cognizance of the study of major religions in the world from the perspectives of comparison of the religions, their ethics and individual philosophies. This study will look at world religions such as Christianity, Islam, Hinduism, Buddhism, Zoroastrianism, Shintoism and more.

In the field of world religion, the main world religions are classified as Abrahamic, Dharmic or Taoic religions. The Abrahamic religion consists of great monotheistic religions such as Christianity, Islam and Judaism which claim Abraham's descent as part of their sacred history. Other religions such as the Bahai faith that fits this description are sometimes included.

Dharmic religions constitute a family of religions that originate from Indian subcontinent. They include Hinduism and three other religions that have spawned from it – Buddhism, Jainism and Sikhism. Zoroastrianism is usually classified as a Dharmic religion, although it also contains elements originating in Abrahamic tradition. The religion of the Vedic period – proto-Iranian time, is the historical predecessor of the Dharmic religions. Vedic religions gradually changed into the historical Dharmic religions.

Taoic religions are religions or religious philosophy that focuses on the East. This forms a large group of religions including Taoism; Confucianism; Shinto; Taoism and Caodaism. In large part of East Asia, Buddhism has taken Taoic features in its course of evolvement.

Approaches in the study of religions:

The approaches used in the study of world religions are, but not limited to: phenomenological, historical and comparative approaches. Phenomenological approach is however given a major attention in this introductory aspect of this work.

Phenomenological approach to the study of religion:

The word phenomenological comes from phenomenon. Phenomenon, going by the interpretation given by Chambers 21th Century Dictionary, 'means a happening perceived through the senses, especially something unusual or scientifically explainable'. Phenomenology is the science of observing or of describing phenomena. The phenomenological approach to the study of religion is, therefore, a scientific approach to the study of religion. It is an academic exercise wherein a religious event or phenomenon is studied through the process of observation, logical analyses and interpretation of experience perceived during the period of study. In this context, the principle of experiential participation is mostly adopted. The scholar therefore gets involved personally in the study and observes every detail of the processes. It could also involve the use of interviews and the administration of questionnaires. The researcher is particularly interested in knowing whether the event of study is interpretable in a way to bring out knowledge which is original or primary to the particular event being studied. When interviews and questionnaires are used to generate data, it would, therefore, be necessary to collate and analyse the data, using simple descriptive method or a more advance analytical tool.

The under listed are very crucial in a study that involves a phenomenological approach: (1) The researcher has to tread gently when studying a religious aspect that has to do with the emotion of others. He should be able to rejoice with those that are dancing and meditate with those that are meditating. If for example, one finds himself in a mosque as a Christian, one's training as a researcher demands that one behave at that moment as a Muslim. In this way, the researcher will be able to get into the proper world view of the particular event that he is studying. (2) The researcher must suspend judgement until the end. As a researcher making use of phenomenological approach as a tool for his work, one must not rush into judging what one considers to be unusual in the religious practice that one is dealing with. Judgment must be suspended until the final analysis. (3) The capacity to see things as a whole, in totality is crucial in the use of phenomenological approach to study religion. Prof Cantwell Smith is very relevant here. He says that a student of comparative religion must be able to present others' religions in the way that when the believers of such religion see it, they will recognize that it is their religion. (4) Commitment is a word that is noteworthy in a phenomenological approach to the study of religions. The word suggests that when it comes to basic facts, attitudes of some fake prophets should not stop one as a Muslim from having feeling for Christianity. Religious commitment is a new phenomenon in the world of religions. Commitment involves the assumption that the individual believers have a free choice in the matter. That the evidence, have been weighed and that the individual has arrived at a decision peculiar to him, freely without cohesion, compulsion or by any external influence. (5) Understanding is another notion that a researcher using phenomenological approach to the study of religion must be conversant with. Understanding presupposes that a certain distance is placed between the student and the phenomenon being studied, and an attempt is being made to get to the root of the phenomenon under investigation (6) In phenomenological investigation, the question of common ground is taken very seriously. What is universal as opposed to what is local and particular in religion?

Ethics and philosophy of the world religions:

The specific ethics and philosophies of the world religion are so important to this work, that the themes were given special attention.

Authors on world religions:

Finally, this work gives recognition to the writings of a selected authors who have written extensively on world religions such as Mircea Eliade, Ninian Smart and others.

Chapter Two: Islam

Formative period of Islam:

To understand Islam, we need to look at the background from which it emerged. In this context, the immediate environment from which it emerged which is Saudi Arabia shall be examined. Most of the things done during Islam formative period are carried into Islam. It is expedient to investigate the formative period of Islam's existence for one to understand Islam as it is being practiced today.

Jehiliyyah

Jehiliyyah could be described as the Period of the ignorance of the worship of Allah. In the words of M. Watt, religion in pre–Islamic Arabia is called tribal humanism. At this period, emphasis was on man and his own power or energy. It is comparable to courage or endurance. It has to do with loyalty or social obligation to one's group. The inhabitants of southern Arabia were mainly Bedouin who have their livelihood in the sale of camel. The central ethical value during this era is based on blood for blood and life for life. There was the worship of Allah at this time but not as we know of in present day Islam. There were also the daughters of Allah – Al-lat, Al-Uzzah and Menat. There were several sacred places and sanctuaries, one of such is Kaaba. At this period were the Halifs whose faith particularly influences Islam. It means people, of monotheistic tendency who sort one God but was not willing to be involved with Christianity or Judaism.

Muhammad, who was the central figure in Islam, was born around AD 570 in the city of Arabia. His father died before his birth. His mother died when he was six and he was raised first by his grandfather and later by his uncle. Mohammad's early background is not well known. Some scholars believe he came from a well-respected family, but this is not certain. At the age of 25, he married a wealthy 40 years old widow named Khadija.

The call:

As Mohammad grew, his views changed. He came to believe in only one God called Allah, a Monotheistic faith. He rejected the idolatrous polytheism of those around him. By the age of forty, the now religious Mohammad had his first vision. These revelations are what were recorded in the Quran. These visions mark the start of Mohammad's prophetic call by Allah. Mohammed received these visions during the following 22 years until his death in AD 632.

The hegira:

The new faith encountered opposition in Mohammad's home town of Mecca. Because of his rejection in Mecca and the ostracism of his views, Muhammad and his followers withdrew to the city known as medina, which means in full "city of the prophet". The hegira, which means "fight", marks the turning point in Islam. All Islamic calendars mark this date, July 16, 622 as their beginning. In his early years in medina, Muhammad was sympathetic to both Jews and Christians, but they rejected him and his teachings. Upon that rejection, Muhammad turned from Jerusalem as the centre of worship of Islam to Mecca, where the famous black stone *Kaaba* was enshrined. Mohammad denounced all the idols which surrounded the *Kaaba* and declared it was a shrine for the one true God Allah.
With this new emphasis on Mecca, Muhammad realized he must soon return to his home. The rejected prophet did return, and in triumph. He conquered the city. Mohammad now made sure of his political and prophetic ascendancy in Arabia. Active opponents near at hand were conquered by the sword. Between the return to Mecca and Muhammad's death, the prophet zealously and militantly

propagated Islam. The new faith, as a result of Muhammad's militancy, spread quickly throughout the area.

Islam immediately after Muhammad's death:

At the period of his death in 632AD, Muhammad had no written will as regards whom to succeed him. After his death therefore, there was leadership tussle. Power struggle developed. Different factions believed their own methods of establishing a successor were better than their rivals. Some believe that the Caliph should be elected by Islamic leadership and some believe that the successor should be hereditary.

The Quran:

The basic scripture for Islamic doctrine is the Quran. Boa described the central place of the Quran in the Islamic faith as well as the supplementary work:

> The Koran is the authoritative scripture of Islam about four-fifth the length of the New Testament. It is divided into 114 *Surahs* (chapters) parts were written by Muhammad, and the rest based on his moral teachings, written by disciples after his death.

Articles and pillars of Faith in Islam

Five articles of Faith:

The five articles of faith are the main doctrines of Islam. All Muslims are expected to believe these tenants:

(1) God: There is only one true God and his name is Allah. Allah is all knowing, all powerful, and the sovereign judge. In essence, Islamic faith is a monotheistic faith.

(2) Angels: The existence of Angels is fundamental to Islamic teachings. Gabriel the leading Angel appeared to Muhammad and was instrumental in delivering of the revelations in the Quran to him.

(3) Scripture: There are four inspired books in the Islamic faith. They are the Torah of Moses, the Psalms of David, the gospel of Jesus Christ and the Quran. (Muslims believe that the former three books have been corrupted by Jews and Christians).

(4) Prophets: In Islam, God has spoken through numerous prophets down through the centuries. The six greatest are Adam, Noah, Abraham, Moses Jesus and Muhammad; they believe that Mohammed is the last and greatest of all God's prophets.

(5) Last days: The last days will be a time of resurrection and judgment. Those who follow and obey Allah and Muhammad will go to Islamic heaven, a place of pleasure. Those who oppose them will go to hell, a place of fire and torment.

Five Pillars of Faith:

(1) **The creed (Kaluna) reads as follows**: "there is no God but Allah, and Muhammad is the prophet of Allah". This creed is the bed rock of Islamic Faith. One must state this aloud publicly in order to become a Muslim. It is repeated constantly by the faithful.

(2) **Prayer (Salat):** prayer as ritual is central to a devout Muslim. This should be five times a day and at certain set times (early in the morning, at noon, at mid-afternoon, after sunset, and before retiring to bed).

(3) **Alms giving (Zakat),** Muhammad, himself an orphan, had a strong desire to help the needy. The alms originally were voluntary, but all Muslims are legally required to give one fortieth of their income for the destitute.

(4) **Fasting (Ramadan),** faithful Muslims fast from sun up to down each day during this holy month. (30 days of fasting)

(5) **The pilgrimage (Haj)**, the pilgrimage is expected of all Muslims at least once in their life time.

Chapter Three: Christianity

The central figure in Christianity is Jesus. The bible calls Him the author and finisher of our faith. He was sent by God to save mankind from sin and to reconcile man back into God. He, therefore, came to raise an assembly of holy people unto God and to lead mankind to the way of salvation. Christianity did not originate from a vacuum. It took its root from the existing religions with some influence from the Hellenistic environment. This study session will examine the central teachings of Christianity as follows:

4.2.1 The Scripture: The Bible is the inspired word of God, the product of holy men who spoke and wrote as they were moved by the holy spirit 2 Timothy 3:16

> All scripture is given by inspiration of God, and is Profitable for doctrine, for reproof, for correction, for instruction in righteousness.

The scripture contained 39 books of the Old Testament and 27 books of the New Testament.

4.2.2 The Godhead:

Our God is one but manifests in three persons – the father, the son and the Holy Spirit. God the father is greater than all: John 14: 28:

> "Ye have heard how I said into you, I go away, and came again unto you, if ye loved me, ye would rejoice, because I said, I go unto the father: for my father is greater that I".

4.2.3 Man, His Fall and Redemption:

Man is a created being, made in the likeness and image of God, but through Adam's transgression and fall sin came into the world. Romans 3:23:

For all have sinned and come short of the glory of God.

Jesus Christ the son of God was manifested to destroy the work of the devil and give His life and shed his blood to redeem and restore man back unto God:

1 John 3:8

He that committeth sin is of the devil, for the devil sinneth from the beginning, for this purpose, the son of God was manifested, that he might destroy the works of the devil.

4.3.1 Eternal Life and New Birth: Man's first step toward salvation is godly sorrow that works out repentance. The new birth is necessary to all men, and when fulfilled produces external life. II Cor. 7:10:

For godly sorrow worketh repentance to salvation not to be repented of: but the sorrow of the world worketh death.

1 John 5:12

He that hath the son hath life, and he that hath not the son of God hath not life.

4.3.2 Water Baptism: Baptism is by immersion. It is a direct commandment of our Lord, and is for those that believe only. The ordinance is a symbol of the Christian's identification with Christ in His death, burial and resurrection.

Matthew 28:19

Go ye therefore, and teach all nations, baptizing them in the name of the father, and the son, and the Holy Ghost.

4.3.3 Baptism in the Holy Ghost: The Baptism in the Holy Ghost and fire is a gift from God as promised by the Lord Jesus Christ to all believers.

Matthew 3:11

I indeed baptize you with water unto repentance: but he that cometh after me is mightier than I, whose shoes I am not worthy to bear. He shall baptize you with the Holy Ghost and with fire.

4.3.4 Sanctification: The bible teaches that without holiness, no man can see the Lord.

Hebrew 12:14

> Follow peace with all men and holiness, without which no man shall see the Lord.

Resurrection of the just and the return of our Lord

Acts 1:11

> Which also said, ye men of Galilee, why stand ye gazing up into heaven? This same Jesus, which is taken up from you into heaven, shall so come in like manner as ye have seen him go into heaven.

4.3.5 Hell and External Retribution: The one who physically dies in his sins without Christ is hopelessly and eternally lost in the lake of fire.

Hebrew 9:27

> And as it is appointed unto men once to die, but after this the judgment.

Chapter Four: A brief comparison of Islam and Christianity

Many of the Islamic beliefs seem to come from the Bible, but yet in spite of the influence and similarities, the difference in the beliefs of the two faiths are striking. This aspect of the study will, therefore, examine some basic concept in both religions.

God in Islam and in Christianity:

Islam teaches the unity of God's essence and personality, explicitly excluding the Trinity as taught in the Bible. Islam has God divorced from His creation, so unified to Himself that He cannot be associated with creation. His transcendence is so great that He acts impersonally. Their doctrine of predestination and the fact that both evil and good came from Allah make their God very capricious. Whatever Allah chooses becomes right, this makes any true standard of righteousness or ethics hard to discern and practically impossible to establish. This is unlike the righteous God of the bible the very word righteous means standard. Righteousness as a concept is well spelt out in Christianity.

Christ as presented by both Islam and Christianity:

In Islam the person and work of Jesus Christ are not seen in the same way as in Christianity. For the Christians, the resurrection of Jesus Christ as the incarnate Son of God is the vital cornerstone of faith, yet the Muslim does not hold either of these truths – that Christ is the Son of God or that He rose from the dead. Islam does believe that Jesus was a sinless prophet although not as great as Muhammad.

Although, Surah 3:45-47 in the Quran speaks of the virgin birth of Christ, it is not the same biblical virgin birth. The idea that Allah had a son is repugnant to them. Christ's claims for His own deity and Son ship are unequivocal. In John 10:30 He claims equality with the father when he states "I and the father are one". Not only is the Son ship of Christ important parse, but the deity of Christ is also an important point of difference between Christianity and Islam since Islam denies the doctrine of the Trinity and of the crucifixion, the Quran states in Surah 4:76. "They slew him not, nor was he crucified, but it appeared so into them". Most Muslims believe Judas was put in the place of Christ, and Christ went to heaven? The bible teaches that Christ went to the cross to pay the penalty for man's sins that He died and was raised from the dead and that He appeared to the disciples and then ascended to heaven.

Sin and Salvation

The Muslim operates under a legalistic system and must earn their salvation by it. They hold the articles of faith and follows the pillars of faith. For the Muslims, sin is lack of obedience to Allah. Man is sinful by act only not by nature.

The Critical Issue of Salvation in Religions

Salvation in Christianity:

The Christian perspective to the issue of salvation is simply that no one can merit the grace of God by the works of the law, or by good deeds or by meditation etc. It is by grace alone. Ephesians 2: 8-9 says:

> For by grace are ye saved through faith, and that not of yourselves:
> it is the gift of God not of works, lest any man should boast.

The fact of faith does not negate the aspect of human's responsibility. The essential ingredient of salvation however remains the grace from God. It is the potential solution to the problem of sin. The Bible assumes that all have sin and come short of the glory of God and that all have been concluded under sin, the initial sin of Adam. It was grace that made God to release His only begotten son, Jesus Christ to go to the cross of Calvary to shed his blood there and die for the remission of Man's sin. Salvation of God must, however, be claimed personally by the sinful man. The recognition of man's sinful state is followed by the acceptance of the atoning sacrifice of Christ. This is what is referred to as salvation.

The effort that humans have to make in order to be saved is to open their heart and receive the gift of God (salvation), through faith. It requires complete trust on God's promises. Faith is the proper human response to God's initiative, not a mere form of positive thinking. In order to accept God's free gift in Christ, humans have to constantly admit their failure to attain salvation by themselves. This situation conflicts sharply with their tendency toward self-justification, with their pride. God wishes that all people should accept his gift of salvation (1Timothy 2:4, 2 Peter 3:9)

> Who will have all men to be saved, and to come into the knowledge of the truth, the Lord is not slack concerning his promises, as some men count slackness; but is long suffering to us word, not willing that any should perish, but that all should come to repentance.

He, however, does not want to force anybody into it. It has to be voluntarily.

Salvation in Islam

Islam teaches that all people are sinners (Quran 16:61) and that salvation can be attained through observing the five pillars of Islamic practice:

(1) The belief that Allah is the only God and that Muhammad is his messenger.

(2) Performing the five daily prayers.

(3) Fasting throughout the month of Ramadan.

(4) Charity, giving alms to the poor.

(5) The pilgrimage to Mecca at least once in a life-time, if one can afford it.

A Muslim hope that by performing these works, at the judgment day, if the recorded good deeds exceed the bad ones, he or she will reach the paradise of material and sensual delight. Faith in Allah and belief that salvation is by His grace is also encouraged. Yet Allah reserves the absolute right to send the deceased to wherever he pleased, paradise or hell. Those who do not submit their lives to the demand of Allah will surely be thrown into hell, a place of extreme physical pain (56:42-45).

Salvation in African traditional religion:

Some people are of the opinion that the hereafter is at a distance place. Such people do bury food and weapons with the deceased. They believe that the food and weapons will serve useful purposes on the journey which the deceased have embarked upon. For some people hereafter is geographically here, being separated by virtue of being invisible to human being. On getting there, the soul would be admitted by older spirits. The Lodagoa believe that the soul must cross a river before getting to his destination. On getting to the other side, the nose is broken so that the departed speak in nasal tones.

An Analytical View of Islam and Christianity:

Islam and Christianity are considered to be world religions. African traditional religion is also recognized across Africa, and in some other parts of the world such as Cuba and Brazil.

Differences:

(1) Islam teaches one Godhead (Allah is one). Christianity teaches unity of the Godhead. Three Gods in one and they co-exist, God the father, God the son and God the Holy Spirit. God the father is the head. African traditional religion teaches an indigenous one Supreme Being who could be reached through the lesser gods, and belief about Him (the Supreme Being) varied across Africa.

(2) Immanence of God is variously taught by the three religious' traditions. Islam teaches that the transcendence of God is so great that he acts impersonally. God is divorced from his creation, so unified to Himself that he cannot be associated with creation. Christianity teaches God's immanence that touches on creation. God is actively present in the world of creation through the third person of the trinity (Holy Spirit). African traditional religion also teaches about God's immanence, that God is present everywhere in the world of creation, especially through the divinities and ancestors.

(3) Islam disagrees with the status of Deity given to Jesus Christ. According to Quran, Jesus is not God's son but a prophet of God in lower status than Mohammad. Christianity teaches the Deity of Christ. In Yoruba mythology *Ela*, a holy deity, shares affinity with Jesus of the Bible.

(4) Islam and Christianity are scriptural religious traditions while African traditional religion is unscriptural religious tradition. The scripture of Islam has the Quran as the foundation and Christianity has the Bible as the foundation, but it is astonishing, the wideness of the difference between the two scriptures.

(5) Islam is both a religion and a cultural identity which cannot be separated from the people. Christianity is essentiality a relationship between man and his God. African traditional religion as a religious tradition is entirely an intrinsic part of the society. You are religious within the circumference of what the society believe as religious norms, ethics, and practices.

(6) Both Islam and Christianity show a lot of divergence in their teaching. Take for example the concept of predestination. Islam believes that both evil and good comes from God. Christianity teaches that God is not the author of evil. African traditional religion teaches that God should be dreaded. There is somehow a similitude between the concept of predestination as taught by Islam and in African traditional religion. As far as righteousness is concerned, what Allah chooses becomes right, this makes any standard of righteousness or ethic hard to discern. This is unlike what the Bible teaches. The very word righteous means "a standard". In Christianity, therefore, there is a clear-cut distinction between what is righteous and what is evil.

Chapter Five: African traditional religion

The Africans as a people are essentially religious. In the words of John Mbiti, religion is the strongest element in the traditional set up. It represents the greatest influence on the thinking and life of the African people at large. In Africa, religion permeates into the whole existence of the people. In the day-to-day discussion, in their names in their attitudes to life generally, you discover that religion is something that cannot be avoided. In the social life of an African man, in his civilization, in his education, in his culture, religion is strongly pronounced. There is no way that religion can be isolated from the people. It is an essential part of their being.

African Traditional Religion is the religion that is indigenous to the Africans. One cannot really trace its origin. It is as old as the African people themselves. It is also a revealed religion like any other religions of the world. It is a religion that has been earlier on misunderstood by many investigators across Europe. As a result, it has been erroneously termed primitive, savage, native and fetish.

Because religion is intrinsically inherent in the nature of an African, and because it permeates all the departments of life, a formal distinction is not easily noticeable between the sacred and the secular as we find it in the other religions like Christianity and Islam. There is no clear-cut distinction between the spiritual and the material areas of life. Anywhere you find an African, his religion is there. At work, at recreation, at ceremonies, at the examination room, religion is involved. As a result of civilization and modern changes, you discover that the African religions cannot remain intact, but at the same time they are not extinct. It readily come to surface during difficult situations, and nowadays most African go back to it in secrets.

One salient fact about traditional religion in Africa is that traditional religion is not primarily for the individuals, but rather for his community of which he is a part. In the traditional African set up, there are no irreligious people. You must go along with the society. You partake in the sacrifices and religious ceremonies. You cannot live an isolated life outside the dictate of the societal norms, beliefs, ideas and concepts. As far as the African religion is concerned, though beliefs and practices very numerous, there is no systematic set of dogmas, creed, theology or liturgy as we find it in orthodox religions like Islam and Christianity. People naturally follow and assimilate whatever religious ideas and practices held or observed by their families and communities. The religious traditions are simply handed down from fore fathers and each generation embraces them with necessary modifications. Therefore, in the thinking of Africans, religion and society are two in separable entities.

The structure of African traditional religion consists of the belief in God; belief in divinities; belief in spirit; belief in the ancestors; the practice of magic and medicine.

Mircea Eliade on African traditional religion:

He talked about the appearance of the sacred. In his book, *Patterns in comparative religion*, Eliade interprets religious data according to different or various kinds of symbols of awareness. In light of his definition of religion as the "appearance of the sacred" humanity has expressed the sacred dimension of life through different modalities. Elide observed that some of these modalities are symbolized by the sky, water, moon, seasons and circles of vegetation. Within each modality is an expression of the nature of life. For example, the transcendence of God is revealed in the infinite, accessibility and creative power of the sky.

Awareness of the sacred is the exact opposite of profaneness of the physical environment in life. What is sacred is what is real. People apprehend the sacred through symbols and myths. According to *Eliade*, symbols, myths and rites, especially, those found in non-literate society represent a real evaluation of human self-awareness in terms of a transcendent ultimate reality. It is a kind of evaluation that is radically different from that expressed by historic everyday situation. This mythical pattern of awareness which Elide calls modality of the sacred can be localized and studied. They have the power that is not derived from a particular historic or social set of circumstances. Rather, they represent a unique dimension of human awareness that takes different forms according to different historic periods or cultural uses.

The structure lasts and extends beyond a particular cultural form and reveals the religious meaning of the particular expression as they are prepared. By comparing the historical expression of religious phenomenon *Eliade* says, the interpreter of religion can write a history of symbols and present religious meaning of religious symbols. Elides efforts to understand religious meaning according to different modalities of sacredness in particular cultural – historic situation shows a deep sensitivity to the fact that religions experience is always formed with a particular cultural context. This fact has become a central concern for several scholars who recognized the inexplicable quality of religion while emphasizing that religious meaning is merely dependent on cultural forces and human capacities. One difficulty many interpreters of religion have is that people (whether scholars or devotes) use the same mode of apprehension for example language, emotion social relation, visual experience and attitudes in both religious and non-religious experiences. How can a scholar distinguish between religious and non-religious elements or how can a person apprehend the religious experience by someone from a different cultural background or religious tradition?

African traditional religion: Comparison with Islam and Christianity

(1) The three religions are institutionalized, whereby Christians worship in Churches and Muslims worship in Mosques and African traditional religion has sacred places of worship with cults for the localized deities and priests attached to the cults. Both Islam and Christianity also have priests; Imam in Islam and Pastors in Christianity. The extent of institutionalisation in African traditional religion is however highly controversial.

(2) Both Islam and Christianity believe in the existence of Angels, while in African traditional religion the belief in spirits is also profound.

(3) All the three religious traditions teach punishment for wrong doings. Both Islam and Christianity teach about heaven for those who lived well and hell for wrong doers. John Mbiti says, concerning African traditional religion there is no belief that a person is punished in the hereafter for what he does wrong in this life. In African religious concept, when punishment comes, it comes in the present life. For that reason, misfortunes may be interpreted as indicating that the sufferer has broken some moral or ritual conducts against God. In African religion however, there is a demarcation between the heaven of the pot-shells and the heaven of the ancestors. The heaven of the pot shells is a place of torment while the heaven of the ancestor is a place of bliss and satisfaction.

(4) All the three religious traditions teach about the Supreme Being and His primordial nature. That God is transcendent, and that He can only be conceptualized in one's mental faculty.

Chapter Six: Common Themes in Islam and Christianity

Our endeavour here is to make a comparative study of some very important theological themes in both Islam and Christianity. It is important for us to note that the two religions are monotheistic; that is, each of them proclaims only one God who is supreme and who is the creator of the heaven and earth. It is also noteworthy to emphasize that the two religions are scriptural religions. The Bible is the official scripture of Christianity while Quran is the official scripture of Islam. These scriptures are believed to be holy and that they represent the authentic words of God to mankind. Another thing which is very crucial for us to note is that both the Holy Bible and the Holy Quran show a great deal of divergence. we shall extract some very major, but common themes in both the Holy Bible and the Holy Quran for the purpose of comparative discourse and analysis.

Who is the Lord God and who is Allah?

The description of God in the two religions portrays some similarities and also divergent. Both the Quran and the Bible give vivid description of both Allah and the Lord God.

It is needful, to start discussion with proper explanation of the concepts of God in the two religious' traditions, Christianity and Islam. We need to know who God is in the two religions. We need to understand the nature and the characteristics of God in the two religions. How comparable is God in the two religions? Can we confidently say that God in the Islamic religion is the same God in the Bible? A comparative analysis of the concept of God will help us to do justice to this question.

Who is Allah in the Holy Quran?

The Quran, **Surah 1** has it that *"in the name of Allah, the beneficent, the merciful. Praise be to Allah Lord of the world."* **Surah** one further ascribes the ownership of the Day of Judgment to Allah. **Surah 7:180** maintains that Allah has many fairer names. That he should be involved by these names.

Surah 112 puts it that Allah is the One, the Only one. He did not bear any one and he was not born by anybody. The opening of the Quran, **Surah 1** should be recited several times in the five daily prayers. It is important to note the description of Allah as beneficent and merciful and also as Lord of the world in the Quran **Surah 1**. Allah is also described in this Surah as the owner of the Day of Judgment, as the one who shows the straight path – that is, the right path. He also shows favour and he could also show anger.

Surah 5:18 also ascribes sovereignty of the heaven and earth and all that is between them to Allah. To the Quran, Allah is absolute, independent, unique and sovereign. He is described as the first and the last. Quran further describes Allah as self-existent and self-sustained. He does not need anything but everything needs him. Nothing happens apart from his will. The knowledge of Allah compasses the whole universe. He always sustains it is important to note the many names of Allah in **Surah 8**. The names are: mighty, wise, best of plotters, seer, helper, hearer, knower, severe in punishment, the forgiver, and the merciful God.

Quran and the Love of Allah

Allah's love for people is conditional. In **Surah 3: 31**, Muhammad says *"If ye love Allah, follow me, Allah will love you and forgive your sins."* This shows that we can only gain Allah's love by first loving him. Allah is merciful and loving toward those who ask pardon from him and then turn to him in total repentance. **Surah 11:90** maintains that first, they must ask for pardon and then, turn to him.

Quran, in **Surah 2: 195, 2: 222,** and in others verses put it that Allah's love is for those who are good. **Surah 85: 14** relates Allah's love to his ability to forgive – that is in the context of those who believe and do good works.

Quran and Bible on the question of Allah's fatherhood:

Islam rejects the concept of God as our father. From the Islamic perspective, it is degrading for God to be seen a father. In the Bible, the Lord God is the supreme being, the Almighty God. In **Exodus 34:6-7,** the Bible reveals the Lord God unto us as a compassionate and gracious God, yet he does not leave the guilty unpunished. The description of the Lord God as compassionate and gracious is also repeated in **Psalm 145:8-9** and in **Psalm 86:15.** The Bible describes God as all knowing he searches and knows us. This is found in **Psalm 139:1** and verses **23** and **24**. The Bible, in **Leviticus 19:2**, describe God as holy (I, the Lord your God, am Holy). He is perfect and just in **Deut 32:4.** The themes of the justice and holiness of God is taken seriously in the Bible. The justice and holiness of God can be propitiated through ransom, or sacrifice. In the Bible, God is seen a father. God as a father figure is well pronounced in the Bible.

The Bible also describes God as infinitely wise. In this regard, **Roman 11:33** exclaims *"Oh, the depth of the riches of the wisdom and knowledge of God"*. In **Deuteronomy 6:4,** God is one. The Lord our God, the Lord is one. God's oneness is likewise expressed in **1 Corinthians 8:6** *"There is but one God, the father, from whom all things came and through whom we live.*

There are more names of the Lord God throughout the Bible. For instance, in **Deuteronomy 10:17**, God is known as *God of gods and Lord of lords*. In **Gen 14:19,**

God is described as *God most high*. In **Gen 14:19, 22** God is described as the creator of heaven and earth. The Lord God is described as *the shield* in **Gen 15:1** and also in **Psalm 3:3.** In **Psalm 18:1** the Lord God is described as strength. In **Isaiah 25:4**, he is described as refuge.

The Bible and the love of God

The Bible has it that God is love. This is found in **1 John 4:8**. This is core to the very essence and the personality of God. God is a loving God. He loved man before man ever responded to his love. In **John 3:16**, the Bible says, *"For God so loved the world that he gave the only begotten son, that whoever believeth in him would not perish, but will have everlasting life."*

God's love is unconditional, unearned and unmerited. **1 John 4:10** say that "*this is love: not that we loved God. But that he loved us and sent his son as the atoning sacrifice for our sins.*"

Similarities of Bible and the Quran on who God is?

1. Both the Bible and the Quran affirms that God is merciful. To the Bible and to the Quran, God is a merciful God.
2. The ownership the heaven and the earth is ascribed to God by both the Quran and the Bible. Allah owns both heaven and the earth, God owns the heaven and the earth. This is so in both the Quran and the Bible.
3. The fact that God and Allah are Gods of justice as affirmed by the Bible and the Quran.
4. The judgment day is held by both the Bible and the Quran. On the judgment day the Lord God and Allah will judge human beings.
5. In both the Quran and the Bible, God is absolute, and sovereign.

6. Both the Bible and the Quran describes Allah and the Lord God as the first and the last
7. Both the Quran and the Bible describes God and Allah as self-existence and self-sustained.
8. In the Quran Allah has many names, likewise, in the Bible the Lord God has many names

Divergence of the Bible and the Quran:

1. In the Quran and in the Bible God is absolute. In the Bible however, in spite of the fact that God is absolute, he relates with his creature. But in the Quran God's absoluteness is without reference to the creature.
2. In the Bible God is seen as father figure, but in the Quran, it is degrading to consider Allah as father.
3. In Quran, the justice and holiness of God cannot be propitiated by ransom or sacrifice, but in the Bible, God can be appeased
4. In the Quran, God's love is conditional, but in the Bible, God's love is unconditional.
5. In the Quran, Allah's love is subject to or conditioned on whether human beings love him or not, but in the Bible, God first love us, even when we are sinners.

The Concept of Trinity in the Bible and in the Quran:

What is the concept of the trinity all about? How does both the Holy Bible and the Holy Quran see and interpret the trinity? Does the Quran believe in the trinity at all? These questions and more constitute the theme of discourse in this module.

Quran:

Quran, **Surah 4:171** says *"o people of the scripture! Do not exaggerate in your religion nor utter aught concerning Allah save the truth. The Messiah, Jesus, son of Mary was only a messenger of Allah, and his word which he conveyed unto Mary, and a spirit from him. So, believe in Allah and his messengers and say not three ... Allah is only one God."*

Quran, **Surah 3:64a** says:

"O the people of the scripture! Come to an agreement between us and you: that we shall worship none but Allah, and that we shall ascribe no partner unto him, and that none of us shall take others for lords beside Allah."

Bible:

Mt 28: 19 says: *"therefore go and make disciples of all nations, baptizing them in the name of the father and of the son and of the Holy Spirit, and teaching them to obey everything I have commanded you. And surely, I am with you always, to the end of the age."*

2 Cor 13:14 says: *"may the grace of the Lord Jesus Christ, and the love of God and the fellowship of the Holy Spirit be with you all."*

John 5:23: says *"that all men should honour the Son, even as they honour the Father. He that does not honour not the Son does not honour not the Father which hath sent him."*

Quran and the Person of Jesus:

- Jesus the Messiah, son of Mary was only a messenger of Allah. The Quran sees Jesus as the Messiah, but not as the son of God.
- It is far away from the transcendent majesty of God that He should have a son.
- Only Allah should be given worship.
- He begetheth not nor was he begotten.
- The statement above is clearly a contradiction to the Biblical statement that Jesus is the only begotten son of God.

Bible on the Person or Deity of Jesus:

Going by the Nicene creed, the Christian belief is in one God, the father Almighty, and in one Lord Jesus Christ, the only begotten son of God and in the Holy Spirit, the Lord and giver of life.

The historic Athanasius creed likewise affirms, the father is Lord, the son is Lord, the Holy Spirit is Lord, and yet there are not three lords but one Lord.
In the Godhead, Jesus is equal in glory with father and the Holy Spirit. Jesus demonstrated his divinity in that even the wind and the leaves obeyed him. See this in **Mathew 8: 27**. His divinity was also demonstrated in his various miracles. See **Luke7:11, 8:41**, and **John 11:43**.

Again, Jesus made claims that only the Lord God, the great "I AM", can make. For example, he claimed he is the bread of life (**John 6.48**). That he is the light of the world (**John 8.12**). In **John 11:25** he said, *"I am the resurrection and the life."*

Looking closely at the Quran and the Bible, one discovers a close affinity in that both Quran and the Bible affirms the messiahship of Jesus Christ. Both see Jesus as one who came from God. The area of divergence is the sonship of God. Quran differs greatly to the Bible in this regard.

Quran denies the deity of Jesus, but the Bible teaches that Jesus is the eternal son of God, the express image of the invisible God.
Both the Quran and the Bible speak of the unity and oneness of God, but they differ greatly in their presentation of who God is. The Quran holds Jesus was only a messenger, through the Messiah, the New Testament affirms that God is one but he is revealed as father, son and the Holy Spirit.

The Concept of God's Will:
Let us start by looking at the various scriptures. What do the scriptures say about the will of God? Both the Bible and the Quran have divergent views on the subject matter. That notwithstanding, a level of similarity could be noticed in the opinion of both scriptures.

Holy Quran:
Surah 9:51 says: *naught befalleth us save that which Allah hath decreed for us. He is our protecting friend. In Allah let believers put their trust.*
Surah 4:79a says *whatever of God befalleth thee (o man) it is from Allah, and whatever of ill befalleth thee it is from thyself.*
Surah 13:27b says: *lo! Allah sendeth whom he will astray, and guideth unto himself all who turn unto him.*

Surah 16:93 says *had Allah willed, he could have made you all one nation but he sendeth whom he will astray and guideth whom he will, and ye will, indeed, be asked of what ye used to do.*

Holy Bible

James 1:13-14 maintains that *when tempted, no one should say, God is tempting me. For God cannot be tempted by evil, nor does he tempt anyone; but each one is tempted when, by his own evil desire, he is dragged away and enticed.*

1 Timothy 2:4 – says *God our saviour wants all men to be saved and to come to knowledge of the truth.*

Eph.1:4-7 has it that, *for He, God, chose us in him (Christ) before the creation of the world to be holy and blameless in his sight. In love he predestined us to be adopted as his sons through Jesus* Christ in accordance with his pleasure and will – to be praised of his glorious grace, *which he has freely given us in the, he One loves. In him we have redemption through his blood, the forgiveness of sins, in accordance with the riches of God's grace.*

The will of Allah and the Lord God:

The Holy Quran

From the Quran, it is deduced that Allah is over all things and that Allah will determine all things. This assertion is documented in the Quran, **Surah 9:51**. Whatever Allah wills happens and whatever Allah does not will does not happen. **Surah 4:79** however clarifies that evil predicament of man is outside the will of Allah – *"whatever of ill that be falleth thee is from thyself."* The Quran holds that Allah determines human path for both good and evil. In this context, **Surah 44:4** holds that Allah sendeth whom he will astray, and guideth whom he will. Allah determines the present and eternal destiny for good or ill of human beings.

Orthodox Islam teaches the absolute predestination of both good and evil – that all our thought, words and deeds whether good and evil, were foreseen, fore-ordained, determined and decreed from all eternity, and that everything that happens takes place according to what has been written for it.

Some in Islam hold that humans have a free will or at least a limited free will, while most hold to the immutable and determinate will of Allah. The Quran maintains that the will of Allah dominates and determines all things. The will of Allah determines one's final destiny.

The Holy Bible

The Bible expounds, in clear terms, the will of the Lord God, in **James 1:13-14.** Already cited earlier, God cannot be tempted by evil, nor does he tempt anyone. Temptation comes when one is drawn by his own evil desire. The Lord God is not the author of evil. 5The Lord God is so kind that he will not allow you to be tempted beyond what you can bear. That assertion is documented in **1 Cor 10:13**. God may permit evil, but that does not mean he is the author of evil. The devil is the author of evil by causing man to sin. The Bible makes it clear that human beings are responsible for their actions. This is because, not only do they inherit the sin of Adam and Eve by birth, they also repeat the same sin of disobedience. Everyone is accountable unto God. God did not predestine anyone unto destruction but when someone persistently hardened his or her heart, God will give him over to judgment. This can be seen in **Rom 1:24,26**. God did not will anyone into damnation, but rather he reacts to humanity's damnation with mercy. If anyone will turn from his evil deeds and embrace the saving grace of God, who is Jesus Christ, God will respond to the person with love. He or she will inherit salvation. This is found in **John 3:18**. **1Timothy 2:4** states clearly that God wants all men to be saved.

Divergence between the Quran and the Bible

The determinative will of Allah is absolute. What he wills will be done. He has the absolute power to lead astray or to guide whom he will.

In the Bible, the Lord God cannot deny his heart of love. His will and heartfelt desire are that all should be saved. That is why God did not spare his only begotten son, but gave him up as a ransom for man's sin. We can see this in **Romans 8:32**.

God will not lead his creature astray as documented about Allah in the Quran.

Similarity

The phrase that God does not tempt anyone in **James 1:13** looks similar to **Surah 4:79a** where the Quran says, "*whatever of ill befalleth you is from thyself not from Allah.*"

Chapter Seven: The person of Jesus in the Quran and in the Bible

The question of who Jesus really is constitutes different levels of discussion in both the Bible and the Quran. Here, the portrait of Jesus in the two sacred texts was carried out and also, a comparative discourse of the personality of Jesus in the two religions was done.

References on the person of Jesus in the Quran

Surah 6:101 – *the originator of the heaven and the earth! How can he have a child, when there is for him no consort, when he created all things and is aware of all things?*

Surah 5:116a *and when Allah saith: o Jesus, son of Mary! Didst they say unto mankind: take me and my mother for two God beside Allah? He saith: be glorified! It was not mine to utter that to which I had no right*

Surah 3:59 – *Lo! The likeness of Jesus with Allah is as the likeness of Adam. He created him of dust, then he said until him be! And he is.*

References on the Person of Jesus in the Bible

Matthew 3:17 – *And a voice from heaven said, this is my son. With whom I am well pleased.*

Matthew 16: 13 When Jesus came to the coast of Caesarea Philippi, he asked his disciples saying ‘whom do men say that I the Son of God am?

Acts 13: 33 God hath fulfilled all the same unto us their children, in that he hath raised up Jesus again; as it is also written in the second psalm, ‘Thou art my Son, this day have I begotten thee.’

John 1: 1-4, 14 – *In the beginning was the word, and the word was with God, and the word was God. He was with God in the beginning. Through him all things were made; without him nothing was made that has been made. In him was life, and that life was the light of men …… the word became flesh and made his dwelling among us. We have seen his glory, the glory of the one and only begotten of the father, full of grace and truth.*

A Closer look at the Quran concerning the Person of Jesus:

A salient question asked by **Surah 6:101** is that, *how can he (Allah) has a child, when there is for him no consort? Allah has no wife; how can we have a son?* To the Quran, Jesus is the son of Mary, but he is not the son of God. This has nothing to do with the high esteem which the Quran hold Jesus. The Quran called Jesus the "slave of Allah".

In **Surah 19:30** the Quran called him the messenger of Allah. In **Surah 19:30**, the Quran categorized Jesus among the prophets. The Quran maintains the messiahship of Jesus Christ, but denies his sonship. The Quran mentioned Jesus or Isa twenty-five time and also mentioned Jesus as the son of Mary twenty-three times.

It is to be noted that Muslims hold the opinion that Jesus and Mary were being worshiped as separate Gods beside Allah. We can see this in **Surah 5:116**. This is a very wrong impression and far from the truth, this is probably due to the exalted position given to Mary by the council of Ephesus in AD431 where-in Mary was described as mother of God. This proclamation does not make Mary to be divine in any way.

A Closer Look at the Bible on the Person of Jesus

In the New Testament, Jesus was called the son of Mary once. This is **in Mark 6:3**. He was called the son of Joseph twice in **Luke 3:23** and **John 6:42**. In the book of Daniel he was called the son of man. This is in **Daniel 7:14**. Jesus is referred to as the son of man or son of God over forty times in the gospel of **Matthew**. He is referred to as the son of God sixty times in the Gospels and the Epistle of John.

In **Matthew 3:17**, God exclaims that Jesus is his son whom he loves.

In **Matthew 14:33**, the disciples worshipped him saying, "truly you are the son of God,

In **Matthew 16:15-16** peter professes that Jesus is the son of the living God.

In **Mark 3:11**, the evil spirit cried out, *"You are the son of God."*

The Bible only sees Mary as a devout, humble, obedient and blessed woman of faith. Mary herself professed in **Luke 1:28** that she is the Lord's servant.

There is a wide level of disparity between how both Quran and the Bible conceived Jesus. Both however conclude that Jesus came from God either as a messenger, a servant, a messiah or as the son of God. The two holy books hold different positions on what Jesus is to God. The Bible says that Jesus is the son of God while the Quran maintains that he is a messenger of Allah.

The mission of Jesus:

The mission of Jesus also constitutes serious debate when a comparative study of both the Bible and the Quran is done. The Quran diverge greatly from the Bible on this subject matter.

The Quran:

The Quran sees Jesus as the slave and prophet of Allah.

In **Surah 19:29-33** the Quran says:

29. Then she pointed to him. They said: how can we talk to one who is in the cradle a young boy?

30. He spake: Lo! I am the slave of Allah. He hath given me the scripture and hath appointed me a prophet

31. And hath made me blessed whosoever I may be hath enjoyed upon me prayer and almsgiving so long as I remain alive

32. And (hath made me) dutiful towards her who bore me, and hath not made me arrogant.

33. Peace on me the day I was born, and the day I die, and the day I shall be raised alive!

The Bible:

The Bible in **Philippians 2:7-8** uses the titles slave/servant and saviour, but in a different perspective:

Phil 2:7-8

7. But made himself of no reputation, and took upon him the form of a servant, and was made in the likeness of man:

8. And being found in fashion as a man, he humbled himself, and became obedient unto death, even the death of the cross

Let us go back to the Quran on the titles: slave and prophet of Allah:

According to the Quran, the child Jesus, from the cradle, declares, lo! Am the slave of Allah. We find this in **Surah 19:30**. Slave of Allah is a common designation applied to Muhammad and all Muslims. The child Jesus, according to Quran, also proclaims that he will be a prophet, he will be dutiful to Mary and he will also be a man of prayer and almsgiving. In **Surah 19:32,** the Quran proclaims that he will not be arrogant. In **Surah 19: 32**, he was born and he will die. According to the Quran, all people must die, and Jesus is no exception but his death could not be by crucifixion. **Surah 19:33** also states that he shall be raised alive. However, this does not mean, he will rise again the third day. The thinking in Quran is that Jesus will be raised on the judgment day.

Let us again Reflect on what the Bible says on Jesus as Slave/Servant and Saviour:

The Bible does not call Jesus a slave. Rather, the Bible says that Jesus took upon himself the very nature of a servant and this is a personal choice and out of a humble heart. It was not determined by God that he would be a servant or slave.

Indeed, in his washing of the disciple's feet, Jesus acts the part of the lowliest slave and he calls all his followers to be the lowliest of slaves to one another. We find this in the Gospel of St. John, chapter 13. He was also described as being obedient to Mary and Joseph in **Luke 2:51**. Jesus is an obedient son who does the will of his father.

The Bible also describes him as a person of prayer to the extent that his disciples implore him, Lord teach us to pray. We can find this in **Luke 11:1** the Bible clearly states that Jesus was put to death on a cross and that on the third day, he was raised by God – This is found in **Matthew 16:21**.

In **Matthew 1:2**, the Bible states that Jesus is not just a great servant whose example is to be followed, but indeed our saviour, for he will save his people from their sins.

The crucifixion accounts of Jesus:

To what extent do these holy documents differ on the question whether Jesus was really crucified or not? The Bible account is completely different from that of the Quran.

Let us start with the Quranic account: the Quran, in **Surah 4:157,** says that *Jesus was not crucified, it only appeared as if he was.*

Surah 4:157

And because of their saying: we slew the Messiah, Jesus the son of Mary, Allah's messenger – They slew him not nor crucified him, but it appeared so unto them, and lo! Those who disagree concerning it are in doubt thereof; they have no knowledge thereof, save pursuit of a conjecture, they slew him not for certain.

In **Surah 4:158**, the Quran maintains *that Allah took him up unto himself. Allah was ever mighty and wise.*

Let us turn to the Biblical account of the crucifixion of Jesus:

Matthew 16:21-23

From that time on, Jesus began to explain to his disciples, that he must go to Jerusalem and suffer many things at the hands of the elders, chief priests and teachers of the law, and that he must be killed and on the third day and be raised to life. Peter took him aside and began to rebuke him. "Never Lord", he said. This shall never happen to you. Jesus turned aside and said to peter, "get behind me, Satan! You are a stumbling block to me; you do not have in mind the things of God, but the things of men.

Matthew 27:31

After they (soldiers) had mocked him, they took off the robe and put his own clothes on him. Then they led him away to crucify him.

Let us have a look at the Quran again:

Does the Quran teach that Jesus was crucified? No! The Quran clearly affirms that Jesus, the Messiah was not crucified, but only appeared so unto them. He was not killed at all. This was because Allah came to the rescue. This is found in the Quran, in Surah 4:157-158. In the opinion of the Quran, Jesus was a great prophet, hence, Allah would not allow him to die on the cross.

There are different schools of thoughts on these issues:

1. Jesus hid while one of his companions died in his place
2. Allah made Judas Iscariot to appear like Jesus and to take his place
3. Simon of Cyrene replaced Jesus before the crucifixion.
4. Satan was substituted and was punished for disobedience

In the interpretation of Helmut Gatje, "The Jews came to an agreement to kill Jesus. Then God informed Jesus that He would raise him up to heaven; so, Jesus said to his disciples: who among you will agree to take a form similar to mine and die (in my place) and be crucified and then go straight to paradise? Apocryphal books such as the Gospel of Barnabas are often cited to corroborate positions of the Quran.

It is a common opinion that Jesus is too great a prophet to be crucified. Dawood has it that literally, He (Judas) was made to resemble Jesus on the cross.

According to the Quran and the teachings of Islam, Jesus was not actually crucified or put to death. The generally accepted Muslim view maintains that Jesus did not die, but Allah took him to himself. Jesus was raised to Allah without death. Islam's holds further that since Jesus did not die, his mission was an unfulfilled one, hence, he has to come back to complete his ministry at the end of the time.

The Biblical narrative about the crucifixion of Jesus:

Prophesies that affirm the coming, death and resurrection of Jesus are common to both the Old and the New Testament books of the Bible.

Jesus himself foretells that he will be killed. This is found in **Matthew 16:21-23**, **Matthew 17:22**, **Mark 8:31** and so on. Before his death, he was indeed mocked and flogged, we can find this in **Matthew 20:19**.

Jesus was no doubt, crucified. On the third day he rose again. The rising Lord showed his hands and feet which still had nail marks. We can see this in **Luke 24:40.** This shows without any doubt that Jesus was killed for our sins and he resurrected and had everlasting victory over death. His atoning sacrifice made a reconciliation of man to God. The crucifixion of Jesus is core to Christian faith without crucifixion, there is no atonement for sin.

Apart from the Biblical account, history confirm the facts of Jesus existence – his birth, his death, and his resurrection. This was confirmed by Josephus, the historian of the Jesus in the *Testimonium Flavianum*.

Paul L Maier also affirms that the fact of the crucifixion of Jesus is substantiated by the discovery of an Arabic manuscript of Josephus' antiquities. The Josephus' account holds that "Pilate condemned him to be crucified, that his disciple reported that he had appeared to them three days after his crucifixion.

The difference between the Quran and the Bible on the crucifixion of Jesus is that Quran denies that Jesus was actually crucified. The Bible authoritatively maintains that Jesus was crucified, that he died and that he was buried. Christ was willing to die for the sins of mankind, not that he was forced to die, and that is the reason that his death was an atonement for man's sins. The crucifixion of Jesus is therefore central to the Christian belief. He was the sacrificial lamb of God. He offered himself as an atonement for the sin of mankind.

Grace and faith:

The question that begs for our response here is that: what is the concept of grace and faith in the Quran, compared with the Bible?

Quran:

Let us start by examining the Quranic opinion on the subject matter. The Quran's position is that there is nothing like atonement by grace or faith.

Surah 3:193 says: *Our Lord! Lo! We have heard a crier calling into faith: "belief ye in your Lord! So, we believed. Our Lord! Therefore, forgive us our sins, and remit from our evil deeds, and make us die the death of the righteous.*

Surah 3:135 says: *And those who, when they do an evil thing or wrong themselves, remember Allah and implore forgiveness for their sins who forgiveth sins save Allah only?*

Surah 3:195b says: *So, these who fled and were driven forth from their homes and suffered damage for my cause, and fought and were slain, verity I shall remit their evil deeds from them and verily I shall bring them into gardens underneath which rivers flow.*

Surah 39:53b says: *Despair not of the mercy of Allah, who forgiveth all sins. Lo! He is the forgiving, the merciful.*

The Bible on the subject, Grace through faith:

Romans 3:23-27 *For all have sin and fall short of the glory of God and are justified by his grace through the redemption that came by Christ Jesus. God presented him as a sacrifice of atonement, through faith in his blood. He did this to demonstrate* *his justice, because in his forbearance he had left the sins committed beforehand unpunished-he did it to demonstrate his justice at the present time, so as to be just and the one who justifies those who have faith in Jesus. Where then, is boasting? It is excluded. On what principle? On that of observing the law? No, but on that of faith.*

Romans 4:13 *It was not through the law that Abraham and his offspring received the promise that he would be heir of the world, but through the righteousness that comes by faith.*

Hebrew 4:15-16 *For we do not have a high priest who is unable to sympathize with our weaknesses, but we have one who has been tempted in every way just as we are – yet was without sin. Let us then approach the throne of grace with confidence, so that we may receive mercy and find grace to help us in our time of need.*

The Quran:

How does forgiveness come, according to the Quran?

In the Quran, forgiveness is not by grace through faith. Forgiveness is without atonement.

In Surah 3:193, Allah as Lord, calls people unto faith: believe ye in your Lord! So, we believed our Lord! Therefore, the plea is that Allah forgive sins. This is found in Surah 3:35. In the opinion of the Quran, forgiveness is a reward for those who ask and do not knowingly repeat the wrong they did.

Quran points out that remission of sins is for those who suffered damage for the cause of Allah, those who fought and were slain. Not only that, they will also be brought to gardens in paradise. Forgiveness is not a gift of unconditional love. It is earned by those who suffered damage for the cause of Allah. Because Allah forgiveth all sins, He is the forgiving and merciful one. We can find this in Surah 39:53.

Quran 3:31 admonishes mankind to love Allah. It reads that, “If ye love Allah, follow me (Muhammad), Allah will love you and forgive you your sins”

Surah 4:124 asserts that one enters paradise if one doth good works and if he is a believer.

Faith and work in Allah assure one of paradise. The concept of becoming righteous before God through grace is foreign to the Quran and Islam. Grace through faith in a saviour is contrary to Islamic teachings. Salvation or admittance to paradise is, therefore through faith in Allah and through works.

The Bible on this same issue:

In the Bible, forgiveness is by grace through faith and atonement. The question is, why is sacrifice necessary for forgiveness?

Romans 3:23 made it plain that, *“all have sinned and fall short of the glory of God*”, so we cannot be justified by our works, or our own efforts. How can sinful human beings meet the demands of the Lord God? There is the need for a mediator who is sinless. Romans 3:20-23, in this perspective, puts it that, we are justified freely by his grace through the redemption that came by Christ Jesus.

As the sacrifices of old were presented as means of atonement, so God presented him a sacrifice of atonement. We can find this in Romans 3: 35. The atonement requires blood Leviticus 17:11 says *“the life is in the blood”*. Through sacrifice, one life was offered in the place of another. Christ offered his life unto God for our sake. This is a sacrifice offered not by human beings but by God himself. Jesus, the victim, or the sacrificial lamb also willingly offered himself in love. We can find this in Hebrew 9:14.

It is clear from the atonement that the father and the son were of one will and one love for mankind. Jesus prayed, when it got to the point that he had to offer himself a living sacrifice that, *"not my will, but yours be done"*. This is in Luke 22:42. The sacrifice made to God with the blood of Jesus demonstrated not only God's great love, but also his justice. This is found in Romans 3:25. God justice for the sin of mankind has to be met and satisfied. God could not just be lenient and overlook sin. Justice in God's holy laws requires that the debt of sin must be paid. God in his underserved love, found a way to meet and satisfy his just requirements. God himself was in Christ reconciling the world to himself.

Angels and Jinn

How do the teachings of both Quran and the Bible differ on angels and jinn (Jinn is the Islamic word for angel)? This question constitutes the theme of discussion here.

Let us start with the Quran:

The Quran holds that all angels prostrated before man at his creation, except the devil.

Surah 38: 71-74 has it that *When thy said unto the angels: Lo! I am about to create a mortal out of mire, and when I have fashioned him and breathed unto him of my spirit. Then fall down before him prostrate, the angels prostrate, every one, saving Iblis (devil), he was scornful and became one of the disbelievers.* The Quran also states that angels pray for believers' forgiveness. We can find this in Surah 40:7. *Those who bear the throne, and all who are round about it, hymn the praises of their Lord and believe in him and ask forgiveness for those who believe (saying) forgive those who repent and follow thy way: ward off from them the punishment of hell!* We find it also in Surah 42:5b that:

The angels hymn the praise of their Lord and ask forgiveness for those on the earth. Lo! Allah is the forgiver, the merciful.

The Quran also posits that angels aid Muslims. We find this in Surah 8:9 as follows: *when ye sought help of your Lord and he answered you (saying): I will help you with a thousand of angels, rank on rank.*

It is also found in Surah 6:61b that *he sendeth guardian over you until when death cometh unto one of you. Our messengers receive him, and they neglect not.*

Bible on the Angels:

What is the opinion of the Bible on the angels? In the book of revelation, the Bible has it that Angels worship and praise God and the Lamb (Jesus). This is in Rev 5:11-12. *Then I looked and heard the voice of many angels, numbering thousands upon thousands and ten thousand times ten thousand. They encircled the throne and the living creatures and the elders. In a loud voice they sang: worthy is the lamb, who was slain, received power and wealth and wisdom and strength and honour and glory and praise.*

The Bible is also of the opinion that the angels serve believers.

This is found in Hebrew 1:14. *"Are not all angels ministering spirits sent to serve those who will inherit salvation?"* The Bible is also emphatic that Jesus will return with angels:

This is found in Matthew 25:31-32 *When the son of man comes in his glory, and all the angels with him, he will sit on his throne in heavenly glory. All nations will be gathered before him, and he will separate the people one from another as a shepherd separates the sheep from the goats.*

Let us examine the two holy scriptures more critically. We will start with the Quran.

In Surah 38:71-74, Allah informed the angels about a special creature he was about to make, the human beings. There was a directive from Allah to the angels that after that Allah breathe life into man, all the angels were to fall down before him prostrate. In the opinion of many Islamic scholars, the angels were not really to fall down to worship Adam, but were to merely bow in honour of the great creation wrought by Allah.

Again, it is an important teaching of the Quran that angels pray for Muslims and ask forgiveness for those who believe. Although it is the belief that angels pray for the believers, they do not in any way stand as mediators for the believers.

In the opinion of the Bible, the angels praise and worship God and the Lamb. The Bible, in the book of Revelation, affirms that thousands upon thousands of angels populate heaven, and they worship God and Jesus all the time.

They worship Jesus that: *Worthy is the lamb, who was slain to Receive power and Wealth and wisdom and strength and Honour and glory and praise!* (Rev 5:11-12).

There is no singular statement in the Bible that angels are to worship the creation of human kind. The angels worship only God (Rev 22:9)

The fact that angels pray for believers is foreign to the Bible. The Bible does not even encourage faith or believe in the angels praying for human beings. Angels do not, in any way, serve as mediator between human beings and God. Angels are powerful but they are not to be prayed to or worshipped.

On the issue of whether angels serve believers or not, the truth of the Bible is that angels are ministering spirits whom God sends to serve the heirs of salvation. This is confirmed in Luke 16:22. There is the case of angels who fed Elijah in **1 King 19:5-6**. The angel of God delivered Daniel from the mouth of lions in the book of **Daniel 6:22**. The angels opened the doors of the jails for imprisoned apostle Paul. This is in **Act 5:19.**

Chapter Eight: Zoroastrianism and Jainism

Zoroastrianism is a religion characterized by salvation through the law. It is not dependent on Judeo Christian or Islam faiths. It could be traced to the priest/ prophet Zoroaster. Zoroaster, according to Greek transcription served as a major stimulus for developments in both Judeo-Christian and Islam Ffaiths. Zoroaster apparently lived in Eastern Iran, but he was dated 6,000 years before Plato, the Parsees themselves, as the followers of Zoroastrianism were called, dates him 258 years prior to Alexander the great. In the opinion of most scholars, this later date holds more truth and he probably lived in the sixth century before Christ.

In the centre of his proclamation was the God *Abura Mazda.* He affirms Him to be the creator of heaven and earth. He is without image and He is the lawgiver of the whole cosmos [...] Loyalty to *Abura Mazda* excluded the worship of other Old Iranian gods. This strict monotheism resembles that of the Israelites' religion. Zoroaster emphasizes the insurmountable opposition between Almighty *Abura Mazda* and *Angra Mainya,* the manifestation of everything evil. Zoroaster deliberately refrains from attributing both spirits, *Abura Mazda* and *Angra Mainya* to the same source or origin. Once these Gods made their initial ethical choices, they separated themselves and the world into spheres of light and that of darkness. The incompatibility of both gods is, for Zoroaster, the basis for ethical demands.

Although the ethical demands remain mostly in the realm of social ethics, fight against evil, as an example, even against maltreatment of cattle. There will be pleasant dwelling in green pastures for those people, including peasants and herdsmen who adhered to the good.

Zoroaster's doctrine is twofold – an eternity of bliss and an eternity of woe allotted respectively to good and evil men in another life beyond the grave. After death, the soul of the deceased has to cross the *Chinvat* Bridge, which stretches over hell, an abyss of molten metal and fire. For the good, the bridge becomes broader and broader for easier transits into heaven where the pious soul will live in eternal joy in paradise. But for the wicked, the bridge grows narrower until it is like the blade of a razor and the soul falls into the abyss of hell, where there will be eternal torment and suffering. There is also some kind of intermediate states for those whose good and bad deeds are in strict balance.

Zoroastrian religion also knows of a judgment at completion of the whole world. Three thousand years after Zoroaster, the *Saoshyant* or saviour will come and bring the present world to its end. The dead will be resurrected and both wicked and good people will have to pass through a flood of molten metal. The good will pass without harm and enter the new world. The wicked will either be purified or burned, and evil spirits will be burned. After this world-wide purification of the present crisis in the last days, *Abura Mazda*'s sovereignty will be complete, and together with him the good, will enjoy a new heaven and a new earth. Of course, not all ideas in the Iranian religion go back to Zoroaster. Some are later developments.
Zoroastrianism portrays, in its eschatological outlook, a confident optimism. The good will ultimately prevails and the bad will be destroyed, or at least, be purified. Zoroaster's ethical demand from its adherents is a constant battle against the evil and the evildoers.

A Comparative Analysis of Islam, Christianity and Zoroastrianism:

1). Christianity is a religion based on grace. Salvation is by grace and not through personal work. God's word says that we are saved by grace through faith in Christ Jesus and not by our own efforts or works (Ephesians 2: 8-9). In Islam and in Zoroastrianism, salvation is by law and works. Quran 4: 124 asserts that one enters paradise if he does good works – salvation is situated within the framework of legalism.

2). Christianity and Zoroastrianism believe in the concept of the saviour whereas, in Islam the prophet Mohammad showed the ummahs the way to Allah. Jesus is the saviour in Christianity and *Saoshyant* is the saviour.

3). Eschatology (belief in the last days) is common to the three religions.

4). Judgement at the end of the age is a common phenomenon in Christianity, Islam and Zoroastrianism.

5). The name of the Supreme Being in Zoroastrianism is *Abura Mazda.* He is simply called God Almighty in Christianity whereas He is called Allah in Islam.

6). Ethical demands remain mostly in the realm of social ethics, fight against evil in the three religions.

7). Both Zoroastrianism and Christianity expounded the theme of the coming saviour, who will come at the end of the age.

Jainism

History has it that Jainism is one of the oldest religions of the world. It is a religion, a way of life and also, a kind of philosophy which has its origin from India. The Jains, that is, the adherents of Jainism follow the teachings of the 24 conquerors who are also known as *Tirthankers*. One of the 24 *Tirthanker*, Lord *Mahavira* lived in 15th century B.C. One of the main characteristics of Jains' beliefs is the **emphasis**

on the immediate consequences of one's behaviours. Jains could be said to be in the minority with the population of about 5 million, but they are very influential. They spread across India and some parts of the United State, West Europe and in many other places. Jains continue to sustain the ancient *Shrama* (ascetic tradition). Jains, though in the minority, they are formidable. They have significantly influenced the religious, ethical, political and economic spheres in India for over two millennia. Jains stresses spiritual independence and equality of all, with particular emphasis on non-violence. According to Jainism, Self-control is vital for attaining omniscience (*kevala Jinana*) and eventually, realization of the soul's time nature. Jains have an ancient tradition of scholarship. The Jains community is the most literate religious community in India.

Jainism has been a major cultural, philosophical, social and political force since the dawn of civilization in Asia, and its ancient influence has been traced beyond the borders of modern India into the Middle Eastern and Mediterranean religions. At various times, Jainism was found all over South Asia including Sri Lanka and the present-day Pakistan and also Bangladesh.

There are two-major sects of Jainism namely–*Digamber* and *Siretamber*. Some historians believe that there was no clear division of these sects of Jainism until the 5th century. The best available information indicates that the Chief Jain Monk, Acharya Bhadrabahu foresaw famine and led about 12,000 *Digamber* followers to Southern India. Twelve years later, they returned to find the *siretamber* sect in 453. The valabbi council compiled and edited traditional *Siretamber* scripture. Most of the differences between the two sects are minor and relatively obscure.

One of the differences is that *Digamber* Jain Monks do not wear clothes because they believe clothes are like other possessions, increasing dependency and desire for material things, and this ultimately lead to sorrow. *Siretamber* Jain Monks wear white seamless clothes and they believe that there is nothing in Jain scripture that condemns wearing clothes. The *Sadhivis* (nuns) of both sects wear white. The differing views arise from differing interpretation of the same holy books.

Beliefs of Jainism:

Jainism is a religion that is not single-handedly founded by a person. The truth is said to have been revealed at different times by different *Tirthanker*– a teacher who has seen the light and who shows the way. A *Tirthanker* is considered to be omniscient, a role model, but not a god. There are 24 *Tirthankers* that are known. However, historical records about only the last two *tirthankers* remained.

Jains believe that every human is responsible for his/her actions and all living beings have an eternal soul, Jiva. This belief seems to be a cardinal teaching of Jainnism. Jains believe all souls are equal. This is because they all possess the potential of being liberated, and attain *Moksha* – the realization of the soul's time nature. *Tirthankers* are role models only because they have attained *Moksha*. Jains insist that one live, think, and act respectfully and honour the spiritual nature of life. Jains view God as the unchanging traits of the pure soul of each living being, chiefly described as infinite knowledge, perception, consciousness and happiness. Jains do not believe in an omnipotent supreme being, creator, but rather, in an eternal universe governed by natural laws.

Jains' scriptures were written over a long period of time. The most cited is the *tattriartha satira* (book of reality) written by the Monk-scholar, umarvati (a.k.a umasuami). The teachings of Jainism are uniformed with great emphasis placed on rational perception, rational knowledge and rational conduct. Compassion for all lives, human and non-human, is central to Jainism. Human life is valued as a unique, rare opportunity to reach enlightenment. To kill any person, no matter what crime he committed, is considered unimaginably abhorrent. It is the only religion that requires Monks and Laity from all its sects and traditions, to be vegetarians.

Some Indian religions have been strongly influenced by Jains and often the majority of the local non- Jain population have also become vegetarian. It has been gathered from history that various sects of Hinduism became vegetarians due to strong Jain influences. Jains are noted for running animal shelter. Jain's stance on nonviolence goes far beyond vegetarianism. They refuse any food obtained with unnecessary cruelty. Dairy foods are in this category. Potatoes, garlic and onions in particular are avoided by Jains. Devout Jains do not eat, drink or travel after sunset, and they prefer to drink water that is first boiled and then cooled to room temperature. Many Jains do not eat green vegetables and root vegetables once a week. The particular day is determined by the lunar calendar and is *Ashtani* (eight day of the lunar month).

Creation and Cosmology:

According to Jain beliefs, the universe was never created, nor will it ever cease to exist. Time is divided into *utsarpinis* (Progressive Time cycle) and *Ausarpini* (Regressive Time cycle). Every *utsarpini* and *Ausarpini* is divided into six equal periods known as *Aras*. During the *utsarpini* half cycle, ethics, progress, happiness, strength, age, body, religion go from the worse condition to the best. During the *Ausarpini* half cycle, these notions deteriorate from the best to the worst. Jain philosophy is based upon eternal universal truths.

Karma Theory

The Jain religion places great emphasis on the theory of karma. Essentially, it means that all *jivas* reap what they sow. A happy or miserable existence is influenced by actions in previous births.

Ethics of Jainism:

Jainism is a religion, a philosophy and way of life deeply imbibed in ethics. The ethical connotation has to do with equality to all and non-violence to every creature. The religion believes in animal shelter and forbid cruelty even to animal. The Jains are mostly vegetarians because they will not take any food obtained through cruelty to animal. The morality of Jainism centres around the law of karma which has to do with reaping what you sow – what one experience, whether good or bad may have to do with the immediate consequence of one's behaviour.

Philosophy of Jainism:

The teachings of Jainism have great emphasis placed on rational perception, rational knowledge and rational conduct. Jains do reflect on what constitute every day experience. A good example here is the killing of animals. The animals too have life. They are living things. If they are living things, why would they be subjected to brutal treatment even with the use of knife? Compassion for all lives, human and non-human, is central to the philosophy of Jainism. Human life is valued as a unique, rare opportunity to reach enlightenment. To kill any person, no matter what crime he committed, is considered unimaginably abhorrent. Probably this is why many states of the world is shifting away from capital punishment – by hanging, lethal injection and with the use of guns. It is the only religion that requires Monks and Laity from all its sects and traditions, to practice vegetarianism as a philosophy of life.

Chapter Nine: Hinduism

Hinduism is one of the oldest of all religious systems; it is also one of the most complex. Hinduism has spawned a variety of sects holding diverse beliefs; therefore, it is difficult to get an accurate picture of Hinduism without consulting many historical books and commentaries.

John Baker observes:

> It is the essence of Hinduism that there are many different ways of looking at or reflecting on a single object, none of which will give view of the whole, but each of which is entirely valid in its own right. Hinduism is tolerant of other religions because Hindus see sameness in all of them.

Hindu Scriptures

The Hindu scriptures were written within 1400 B.C. – A.D. 500, a period of about 2,000 years and they are very voluminous. They reflect the practices and beliefs which arose during the different periods of Hindu history. The word ***Veda*** literally means wisdom or knowledge. It is the term applied to the oldest of the Hindu scriptures, originally transmitted orally and subsequently preserved in written form. The Veda contains hymns, prayers and ritual texts composed during a period of two thousand years, beginning around1400 BC.

The ***Upanishads*** are a collection of speculative treatises. They were composed during the period 800-600 B.C., and 108 of them are still in existence. The word *Upanishad* conveys the idea of secret teaching. Its treatises mark a definite change in emphasis from the sacrificial hymns and magic formulas in the *Vedas* to the mystical ideas about man and the universe, especially the eternal Brahman which is the basis of all reality. The Upanishad reportedly had an influence upon Gautama

Buddha, the founder of Buddhism, as can be observed in some basic similarities between the *Upanishads* and the teachings of *Mahayana* Buddhism.

The ***Ramayana*** is one of the two epic tales of India, the other being the *Mahabharata*. Authorship is ascribed to the sage-poet Valmiki. The work consists of 24,000 couplets based upon the life of Rama, a righteous king who was supposedly an incarnation of the God Vishnu. The ***Mahabharan*** is the second epic; an immense story of the deeds of Aryan clan. It consists of some 100,000 verses and was composed over an 800-year period beginning around 400 B.C. Contained within this work is a great classic, the ***Bhagavad Gita***, or the "Song of the Blessed Lord". The *Bhagavad Gita* is not only the most sacred book of the Hindus, but it is also the best known and the most read of all India works in the entire world, despite the fact it was added late to the *Mahabharata*, sometime in the first century A.D.

This story revolves around man's duty, which, if carried out, will bring nothing but sorrow. The significance this story has on Hindu beliefs is its endorsement of *bhakti*, or devotion to a particular god, as a means of salvation, since *Arjuna*, the story's main character, decides to put his devotion to Vishnu above his own personal desires. The *Gita* ends with *Arjuna* devoted to Vishnu and ready to kill his relatives in battle. This poem has inspired millions of Hindus who, because of their own situation, have identified with *Arjuna* and his dilemma. The poem offers hope, through the way of devotion, to all people no matter their caste or sex. The poor and the downtrodden who could not achieve salvation through the way of devotion take solace in the poem. These two epic stories, the *Ramayana* and the *Mahabharata*, depict characters that have become ideals for the people of India in terms of moral and social behaviours. The ***Puranas*** are important sources for the understanding of Hinduism. They include legends of gods, goddesses, demons and ancestors, and they describe pilgrimages and rituals to demonstrate the importance of *bhakti*, caste and dharma.

Hindu Teachings

Moksha (*mukti*) is the Hindu term used for the liberation of the soul from the wheel of karma. For the Hindus, the chief aim of existence is to be freed from ***samsara***– the binding life-cycle and the wheel of karma with its endless cycle of births, deaths and rebirths. When one achieves this liberation, he enters into a state of fullness or completion.

The word ***karma*** literally means action and refers to a person's actions and the consequences thereof. In Hinduism, one's present state of existence is determined by his performance in the previous life time. The law of *karma* is the law of moral consequence or the effect of any action upon the performer in a past, present or even future existence. As one performs righteous acts, he moves toward liberation from the cycle of successive births, deaths and rebirths. Contrariwise, if one's deeds are evil, he will move further from liberation. The determining factor is one's *karma*. The cycle of births, deaths and rebirths could be endless.

Samsara refers to transmigration or rebirth. It is the passing through of a succession of lives based upon the direct reward or penalty of one's karma. This continuous chain consists of suffering from the results of acts of ignorance or sin in past lives. At each successive rebirth, the soul, which the Hindus consider to be external, moves from one body to another and carries with it the karma from its previous existence. The rebirth may be to a higher form – a member of a higher caste or god, or down the social ladder to a lower caste or animal, since the wheel of karma applies to both man and animals.

The caste system is a unique feature of the Hindu religion. The account of its origin is an interesting story. Brahma created Manu, the first man. From Manu came the four different types of people, as the creator Brahma has determined. From Manu's head came the Brahmins, the best and most holy people. Out of Manu's hands came the *Kshatriyas*, the rulers and warriors. The craftsmen came from his thighs and are called *Viasyas*. The remainders of the people came from Manu's feet and are known as *Sudras*.

The caste system became more complicated as time went on, with literally thousands of sub castes coming into existence. Today the caste system is still an integral part of the social order of India, even though it has been outlawed by the Indian government. From early times the Hindus revered cow and considered it a possessor of great power. The following verses from the **Atharva Veda** praise the cow, identifying it with the entire visible universe:

> Worship to thee, springing to life, worship to thee when born! Worship, O Cow, to thy tail-hair, and to thy hooves, and to thy form... The Cow is Heaven, the Cow is Earth, and the Cow is Vishnu, Lord of Life. He who hath given a Cow unto the Brahmans wins all the worlds ... Both Gods and mortal men depend for life and being on the Cow. She hath become this universe; all that sun surveys is she (Atharva Veda X: 10).

Ethics of Hinduism:

The sources of ethics of Hinduism includes:

a. ***Moksha,*** this is the Hindu term used for the liberation of the soul from the wheel of karma – for one to be freed from ***samsara*** – the binding life-cycle and the wheel of karma with its endless cycle of births, deaths and rebirths. Liberation from the wheel of karma is necessary for one to enters into a state of fullness. The concept of moksha suggests a disposition to living right and behaving right in a conventionally accepted manner for one to break away from the binding life circle of birth, death, rebirth and death. If one is disposed to a rightful living, one will be freed from samsara and thereby achieve the state of fullness.

b. *karma* suggests one's actions and the consequences thereof. To the Hindus, one's actions in the previous life time is binding on one's present existence. The law of *karma* is the law of moral consequence or the effect of any action upon the performer in a past, present or even future existence. The criteria to move toward liberation from the cycle of successive births, deaths and rebirths is the performance of righteous acts. The determining factor is one's *karma*. The cycle of births, deaths and rebirths could be endless.

c. The caste system is unique as a feature of the Hindu religion. The origin is an interesting story. The first man, Manu, was created by Brahma. From Manu came the four different types of people. From Manu's head came the Brahmins, the best and most holy people. Out of Manu's hands came the *Kshatriyas*, the rulers and warriors. The craftsmen came from his thighs and are called *Viasyas*. The remainders of the people came from Manu's feet and are known as *Sudras*. One, after death, will reincarnate to the higher level of the caste system or the lower level, depending on one's action.

Philosophy of Hinduism:

Reflection, a major approach of philosophy, is vividly observed in Hinduism. The Hindus believe that there are many ways of reflecting on a single object, none of which gives view of the whole but valid in its own right. The philosophy or thinking here is that if one's religion appear far from been perfect and yet it represents for one a symbol of the right part and a similar conviction seems to be the case for others; if one cannot dismiss the religious claim of the others nor assimilate it completely into one's own, a plausible alternative is to assume that all are different creeds, which in spite of meandering and crossing, actually run parallel to meet only at the ultimate end. In this panorama, Hinduism seems to be tolerant of other religions. This philosophy of tolerance makes Hinduism to be more accommodating of other religions.

Further, in the philosophical world-view of Hinduism, the concepts of Moksha and Karma are thought provoking. Moksha is the liberation of the soul from the wheel of karma. Karma has to do with individual actions or behaviours. The philosophy here is that whatever one does has a far-reaching implication on one's success or failure now or in the future; it even could extend to another existence in the circle of birth, death and rebirth. This philosophy upholds the concept of reincarnation and when one reincarnate, the consequences of one's action in the previous existence follows one into the new life. This Hinduism philosophy further on the concept of fatalism. Fatalism upholds the idea that whatever is one's lot on earth should be accepted whole-heartedly. It could be the consequence of one's deeds in the previous existence.

Jainism and Hinduism: A Comparative Reflection:

1). Both Jainism and Hinduism, could be classified as religion, way of life and philosophy of life.

2). The two religions have no particular founders. In Jainism, truth is said to be revealed at different times by different tirthankers – a teacher who has seen the light. Both believe in the worship of gods. In Hinduism, worship to cow is noticed: The following verses in **Atharva Veda** is hereby quoted:

> Worship to thee, springing to life, worship to thee when born! Worship, O Cow, to thy tail-hair, and to thy hooves, and to thy form... The Cow is Heaven, the Cow is Earth, and the Cow is Vishnu, Lord of Life.

3). In the two religions, emphasis is placed on the immediate consequences of one's actions. The both believe in the law of karma. Consequence of one's behaviours. Whatever one sow, one will reap it.

4). The two are not just religions, but also represent the worshippers' way of life and philosophy.

5). The followers of both religions are mostly vegetarians.

6). Jains could be said to be in the minority, while the Hindus were more in population

7). Jains belong the elite, literate stratum of the society while Hindus cut across every sector of the society.

8). Both religions are ethical religions. They believe in good conducts and also in the consequences of bad conduct.

Chapter Ten: Buddhism

Buddhism emerged about five hundred years before the birth of Jesus Christ. The people living at that time had become disillusioned with certain beliefs of Hinduism including the caste system, which had grown extremely complex. The number of outcasts (those who did not belong to any particular caste) was continuing to grow. Again, the Hindu belief of an endless cycle of births, deaths and rebirths was viewed with dread. Consequently, the people turned to a variety of beliefs, including the worship of animals, to satisfy their spiritual vacuum. Many different sects of Buddhism arose, the most successful being that of Buddhism, which denies the authority of the *Vedas*.

The History of Buddha:

Buddhism, unlike Hinduism, can point to a specific founder, Siddhartha Gautama by name. He is also known in history as Buddha. However, in Buddhism, like so many other religions, fanciful stories arose concerning events in the life of the founder. Siddhartha Gautama was born in the fifth century B.C. **The Buddha or the enlightened one** was born about 560 B.C. in north-eastern India. His family name was Gautama; his name was Siddhartha. Siddhartha was the son of a rajah, or a ruler. His mother died when he was just a week old and Siddhartha was cared for by his mother's sister, who was also the rajah's second wife. There was a prophecy given at the time of his birth by a sage at his father's court.

The prophecy said that the child would be a great king if he stays at home, but if he decides to leave home, he would become a saviour for mankind. This bothered his father, for he wanted his son to succeed him as king. Therefore, to keep him at home, his father surrounded him with wealth and pleasures and kept all painful and ugly things out of his sight. Siddhartha eventually married and had a son but was still confined to the palace and its pleasures. One day he informed his father that he wished to see the world outside the palace. This excursion would, forever, change his life for it was during this journey that he saw "the four passing sights" which motivated or informed his decision to go and seek for enlightenment.

Although his father ordered the streets to be cleansed and decorated and all elderly or infirmed people to stay inside, there were those who did not get the message. The first troubling sight that Siddhartha saw was that of a decrepit old man. When Siddhartha asked what happened to this man, he was told that the man was old, as everyone someday would become. Later, he met a sick man and was told that all people were liable to be sick and suffer pain like that individual.

He then saw a funeral procession with a corpse on its way to cremation, the followers weeping bitterly. When he asked what that meant, the prince was informed that it was the way of life, for sooner or later both prince and pauper would have to die. The last sight was that of a monk begging for his food. The tranquil look on the beggar's face convinced Siddhartha that the beggar's type of life was for him. Immediately he left his family and the palace in search of enlightenment. The night he left his home to seek for enlightenment became known in Buddhism as the Great Renunciation.

The former prince, now a beggar, spent his time wandering from place to place seeking wisdom. Unsatisfied by the truths taught in the Hindu scriptures, he became discouraged but continued on his quest. He tried asceticism but this gave him no peace. The fateful day in his life came while he was meditating beneath a fig tree. Deep in meditation, he reached the highest degree of God-consciousness, known as

nirvana. He supposedly stayed under the fig tree for seven days. After that, the fig tree was called the *bodhi*, or the *bo* tree, the tree of wisdom. The truths he learned he would now impart to the world, no longer as Siddhartha Gautama, but as the Buddha, the enlightened one.

When Buddha emerged from his experience under the bo tree, he met with five monks who had been his companions. It was to these monks that the Buddha began his teaching ministry with the sermon at Benares. The sermon contained the following:

Buddha's Sermon at Benares

The sermon is based on two extremes. These two extremes are not to be practices by monks. What are the two extremes? 1. Passions and luxury, which is low, vulgar, common, ignoble, and useless; 2. Self-torture, which is painful, ignoble and useless. By avoiding these two extremes, the Blessed one has gained the enlightenment of the Middle Path, which produces insight and knowledge, and leads to calm, to higher knowledge, enlightenment, *nirvana*.

What is the Middle Path?

It is the noble eightfold Path namely: (1) right view (2) right intention (3) right speech (4) right action (5) right livelihood (6) right effort (7) right mindfulness (8) right concentration. After the sermon at Benares, Buddha started to spread his teachings to the people of India. The India people, disillusioned with Hinduism, listened intently to this new doctrine. By the time of Buddha's death, at age 80, his teachings had become a strong force in India. Sometime after his death, the Buddha was deified by some of his followers, even though veneration of the Buddha is against the basic teachings of Buddha himself.

The Teachings of Buddhism:

A key concept in Buddhism is *nirvana*, the final goal for the Buddhists. Donald K. Swearer gives insight into this important concept. Nirvana has been a troublesome idea for the students of Buddhism. *Nirvana* is a negative term. Literally, it means the "blowing out" of the flame of desire, the negation of suffering (*dukkha*). This implies that *nirvana* is not to be thought of as a place but as a total reorientation or state of being realized as a consequence of the extinction of blinding and binding attachments. There are five precepts taught by Buddhism that all Buddhists should follow:

1. Kill no living thing (including insects)
2. Do not steal
3. Do not commit adultery
4. Tell no lies
5. Do not drink intoxicants or take drugs.

There are other precepts that apply only to monks and nuns. These include:

6. Eat moderately and only at the appointed time.
7. Avoid that which excises the senses.
8. Do not wear adornments (including perfume)
9. Do not sleep in luxurious beds
10. Accept no silver or gold.

Ethics of Buddhism:

The teachings of Buddhism give great insight into the ethical connotation of the religion.

Nirvana is a key concept in Buddhism, it represents the final goal for the Buddhists. *Nirvana* is a negative term. Literally, it means the "blowing out" of the flame of desire, a total reorientation or state of being realized as a consequence of the extinction of blinding and binding attachments of materialistic inclination. The following precepts constitute the subject of morality in Buddhism: Kill no living thing; do not steal; do not commit adultery; tell no lies; do not drink intoxicants or take drugs; eat moderately; avoid that which excises the senses; do not wear adornments; do not sleep in luxurious beds; accept no silver or gold. The morality of Buddhism advances greatly, the concepts moderation and simplicity.

Philosophy of Buddhism:

Buddhism is more of a philosophy and a way of life than religion. It is more or less a philosophy of right living, moderation and abstinence from unnecessary excesses. Central to Buddhism is a human founder who himself has no personal claim to Divinity, but was only deified by his followers after his death. Nirvana itself, which is a core teaching of Buddhism is a philosophy of moderation and avoidance of material acquisition.

Chapter Eleven: Confucianism and Shintoism

Confucianism

Confucianism, as a religion has had a monumental impact on the life, social and political structures of China. The founding of the religion goes back to one man, known as Confucius who was born half-millennium years before Christ.

The Life of Confucius

Although Confucius occupies a hallowed place in Chinese tradition, little is verifiable about this man. The best source available is *The Analects,* the collection of his sayings made by his followers. Long after his death much biographical detail on his life surfaced, but most of these materials are of questionable historical value. However, there are some basic facts that can be accepted reasonably to give an outline of his life.

Confucius was born in Chiu King in China. He was the youngest of eleven children. His birth took place around 550 B.C., in the principality of Lu, which is located in present – day Shantung. He was a contemporary of the Buddha (although they probably never met) and lived immediately before Socrates and Plato. Nothing is known for certain concerning his ancestors except the fact that his surroundings were humble. Confucius revealed about his humble beginning as follows: "When I was young, I was without rank and in humble circumstances".

His father died soon after his birth, leaving his upbringing to his mother. During his youth, Confucius participated in a variety of activities, including hunting and fishing; but, on reaching the age of 15, he said he bent his mind to learning". He held a minor government post as a collector of taxes before he reaches the age of 20. It was at this time that Confucius married. However, this marriage was short-lived, ending in divorce after producing a son and a daughter. He became a teacher in his early twenties, and that proved to be his calling in life. His ability as a teacher became apparent and his fame spread rapidly, attracting a strong core of disciples. Many were attracted by his wisdom. He believed that society would not be changed unless he occupied a public office where he could put his theories into practice. This explains why he held government posts at various levels for many years.

Confucius held minor posts until age 50, when he became a high official in Lu. His moral reforms achieved an immediate success, but he soon had a falling out with his superiors and subsequently resigned his post. Confucius spent the next thirteen years wandering from state to state, attempting to implement his political and social reforms. He devoted the last five years of his life to writing and editing what have become Confucian classics. He died in Chufou, Shantung, in 479 B.C, having established himself as the most important teacher in Chinese culture. His disciples referred to him as King Fu-tzu or Kung the Master, which has been Latinized into Confucius.

China before Confucius:

It is important to understand life in China at the time of Confucius in order to have a better appreciation of the reforms he was attempting to institute. The age in which Confucius lived was characterized by social anarchy. Huston Smith gives insight into the condition of China during this difficult period:

> Instead of nobly holding their prisoners for ransom, conquerors put them to death in mass executions. Soldiers were paid upon presenting the several heads of their enemies. Whole populations unlucky enough to be captured were beheaded, including women, children, and the aged.

> We read of mass slaughters of 60,000, 80,000, 82,000 and even 400,000. There were accounts of the conquered being thrown into boiling caldrons and their relatives forced to drink the human soup. It is easy to see how the need arose for someone like Confucius to provide answers as to how the people could live together harmoniously.

Confucius believed China could be saved if the people would seek for the good of others which was a practice of their ancestors. The role Confucius would play was not as a saviour or messiah but as one who would put the people back in touch with the ancients: "I transmit but do not create. I believe in and love the ancients".

The Veneration of Confucius:

Like many great religious leaders, Confucius was eventually deified by his followers. The following chart traces the progress which led to his ultimate deification:

B.C.195	The Emperor of China offered animal sacrifices at the tomb of Confucius.
A.D.1	He was given the imperial title "Duke Ni, All-complete and Illustrious".
A.D.57	Regular sacrifice to Confucius was ordered at the imperial and provincial colleges
A.D.89	He was raised to the higher imperial rank of "*Eral*".
A.D.267	More elaborate animal sacrifices to Confucius, four times yearly, was decreed.
A.D.492	He was canonized as "The Venerable, the Accomplished Sage".
A.D.555	Separate temples for the worship of Confucius were ordered at the capital of every prefecture in China.
A.D.740	The statue of Confucius was moved from the side to the centre of the Imperial College, to stand with the historic kings of China.
A.D 1068-1086	Confucius was raised to the full rank of Emperor.

A.D.1906 (December 31) He was raised to the rank of Co-Assessor with the deities of heaven and earth.

A.D.1914 The worship of Confucius was continued by the first President of the Republic of China, Yuan Shi Kai.

Ethics in Confucianism:

For one to appreciate the connotation of ethics in Confucianism, there is the need for a background study of life in China before Confucius was born. It was more or less a state of anarch then. It was a period of inter-tribal wars wherein the victors take joy in beheading their victims. Mass execution and burial of the enemies were recorded in the Chinese history. This was a period that Confucius emerged as a great teacher in China. The teaching of Confucius was, therefore, geared towards moral reformation in China. He was, therefore, an advocate of moral reform – to return China to the days of the ancestors when there was peace and love; wherein tolerance as a virtue, was greatly promoted. Confucianism is, therefore, a religious and a philosophical way of life that seek for the good of others. The ethical codes in Confucianism therefore hinge on right way of living, respect for the aged, tolerance of others, love and seeking for the atmosphere of peace in the Chinese communities.

Philosophy of Confucianism:

Confucianism is not a religion in the sense of man relating to the Almighty but is rather an ethical and philosophical system teaching man how to get along with his fellow man. However, Confucius did make some comments on the supernatural which give insight into how he viewed life, death, heaven, etc. He once said, "Absorption in the study of the supernatural is most harmful". When asked about the subject of death, he had this to say: "Chi-lu asked how the spirits of the dead and the gods should be served. The master said, 'You are not able to serve man. How can you serve the Spirits'?

'May I ask you about death?' You do not understand even life. How can you understand death?'" A major philosophy of Confucius is the thinking that his theories could be well-advanced by participating in the politics of his time. with this philosophy, Confucius believed that if he was able to hold government power, he will be able to influence the Chinese people across the social stratum. He also held the philosophy that good life exists in political and social reforms; that with good governance and societal peace, life will be pleasant for everyone. The philosophy of Confucianism, is therefore, to seek the good of others like in the days of their ancestors.

Buddhism and Confucianism: A comparison:

1). The two religions have definite founders: Buddhism was founded by Buddha while Confucianism was founded by Confucius.

2). In the two religions, emphasis is not laid on heaven, but rather to live well and relate well here on earth.

3). The two religions are more of ways of life and Philosophy than being a religion.

4). The two religions are more or less advocate of moral reforms, especially, Confucianism.

5). Confucius, the founder of Confucianism was a deified god while Buddha, the founder of Buddhism was an enlightened one who merely leads others to the right part.

Shintoism

Shinto is the national religion of Japan. It is one of the oldest of all the world's religions. It is unlike other religions in as much as it is basically not a system of beliefs. It has been variously defined. Shinto is purely a Japanese religion, the origin of which is buried in antiquity. The Japanese name for their country is Nippon, which means "sun origin". Until the end of World War II, Japanese children were taught at school that the emperors were descendants of the sun, *Amaterasu. Amaterasu* had allegedly given the imperial house the divine right to rule. In 1946, in a radio broadcast to the Japanese people, Emperor Hirohito repudiated his divine right to rule. They "are honest and upright of heart, and are not given to useless theorizing and falsehoods like other nations". Japanese Emperor Meiji established Shinto as the official religion of Japan in place of Buddhism. However, since the people continued to embrace both religions, in 1877 Buddhism was allowed to be practiced by the people, with total religious liberty granted two years afterward.

Meaning of Shinto

The word Shinto comes from the Chinese word *Shen-tao,* which means "the way of the gods". This term was not applied to the religion until the sixth century A.D., when it became necessary in order to distinguish it from Buddhism. A major feature of Shinto is the notion of *Kami. Kami* is a difficult term to define, but precisely, it refers basically to the concept of sacred power in both animate and inanimate objects.

Ninian Smart elaborates upon the idea of *Kami* in the following manner:

> Shintoism displayed, and still displays, a powerful sense of the presence of gods and spirits in nature. These spirits are called *Kami,* literally, "superior beings", and it is appropriate to venerate them. The *Kami* are too numerous to lend themselves to a systematic ordering or stable

> hierarchy, but among them is the sun goddess, *Amaterasu. Amaterasu* has long held a central place in Shinto belief.

Shinto's history can be divided into a number of stages. The first period was from prehistoric time to A.D. 522 when Shinto reigned supreme among the people of Japan without any serious competition. In A.D. 522, Buddhism started gaining popularity among the Japanese people. In the year A.D 645, The Emperor Kotoku embraced Buddhism and rejected Shinto. From A.D 800 to A.D. 1700, Shinto combined with other religions, mixing with both Buddhism and Confucianism and forming what is called *Ryobu Shinto,* or dual- aspect Shinto. Shinto, by itself, experienced a considerable decline during this period.

Around A.D.1700 Shinto experienced a revival when the study of archaic Japanese texts was reinstituted. One of the most learned Shinto scholars of the period was Hirata, who writes:

> The two fundamental doctrines are: Japan is the country of the gods, and her inhabitants are the descendants of the gods. Between the Japanese people and the Chinese, Hindus, Russians, Dutch, Siamese, Cambodians and other nations of the world there is a difference of kind, rather than of degree. The Mikado is the true Son of Heaven, who is entitled to reign over the four seas and the ten-thousand countries.

Sacred Books

Although Shinto does not consider any one volume as the wholly inspired revelation on which its religion is based, two books are considered sacred and have done much to influence the beliefs of the Japanese people. The works are *ko-ji-ki,* the "records of ancient matters," and *Nihon-gi,* the "chronicles of Japan." They were both composed around A.D.720 and because they report events occurring some 1300 years earlier in the history of Japan, they are considered late works.

Worship:

The basic place for worship in Shinto is at one of the numerous shrines in the country of Japan. Although many Shintoists have built altars in their homes, the centre of worship is the local shrine. Since Shinto has a large number of deities, a systematic worship of all deities is impossible. The Shinto religious books acknowledge that only a few deities are consistently worshipped, the chief being the sun-goddess, *Amaterasu*.

The fact that the highest object of worship from whom the divine ancestors arose is a female rather than a male deity makes Shinto unique among the larger world religions.

Selected authors on world religions:

Chapter Twelve: Selected authors of World religions

Ninian Smart

He is the author of many books. He is the major contemporary exponent of phenomenological approach to religion. His approach is within the tradition of philosophical phenomenology. He was the exponent of phenomenological school of religion. One definition Smart gives is that a religion or the religion of a group is a set of institutionalized rituals identified with a tradition and expressing or evoking sacred sentiments directed at trans-human focus, seen in the context of the human phenomenological environment and at least, partially described by myths and doctrines.

It is important to look at the component of the definition a little more closely as it revealed within a very small space, the essentials of Smart's approach to the study of religion which has had a profound effect on the modern approach to religion. In the first place, Smart poses that religion ordinarily can be very misleading. We need to speak of the religion of a group, identified with a tradition. In this, smart stresses that people who are not simply religious in any general sort of ways are certainly not comparatively religious, they are religious in so far as they belong specifically and exclusively to a particular tradition. Consequently, if we wish to study religion, we need to do so in a way in which people actually experience it. Thus, this basic unit of study must be a specific religious system.

Moreover, religions do not simply spring up out of nowhere: **Any given religion at any given point in time is part of a developing changing tradition**. The history of any religion or denomination or set is an important aspect in the study of it. In the second place, smart holds that any given religion is an internally complex system. A system implied dynamic and structural relation between component parts. Those parts smart call dimensions in any given religious system (Doctrinal, mythological, ethical, ritual, experiential and social) and expressing the relation between them. By focusing attention both on the six characteristic dimensions and the relation between them, smart suggest that religions have their own logical internal structure.

Raimundo Panikkar:

He was born into two major traditions – Catholic Christian and Hindu traditions. Panikkar concerned himself, since his early years, with the harmony of a pluralistic religious world. He holds double doctorate in chemistry/ philosophy and theology. He is the author of many books and articles. He was for many years, a professor of religious studies at the University of California. We shall be concerned with Panikkar's writing on religious pluralism and dialogue, especially inter- religious dialogues. Panikkar, at the beginning of his writings informed us that he would like to propose attitude and model for the proper rhetoric in the meeting of religious traditions. He suggests three attitudes:

a. **Exclusivism**: A believing member of a religion in one way or another considers his religion to be true. Now the claim to truth has a certain built-in claim to exclusivity. If a given statement is true, its contradiction cannot also be true. Again, if a certain human tradition claims to offer a universal context for truth, anything contrary to that universal truth will have to be declared false. If, for example, Islam embodies the truth, truth cannot exist in other religions. This attitude, according to Panikkar has a certain element of heroism in it. You concentrate your life and dedicate your entire existence to something which is really worth of being called a human course, to something that claim not just a partial and imperfect truth, but a universal and even absolute truth. On the other hand, this attitude presents its difficulties. First, it carries with it the obvious danger of intolerance. Truth is many facets and even if you assume that God speaks in an exclusive language, everything depends on one's understanding of it. So that one may never really know the truth.

b. **Inclusivism**: In the present world context, one can hardly fail to discover the positive and true value even of the highest order outside one's own tradition. Traditional religion has to face this challenge. For splendid isolation is no longer possible. The most plausible condition for the claim to truth of one's own tradition is to affirm at the same time that it includes at different levels all that there is of truth where ever it exists.

c. **Parallelism**: If one's religion appear far from been perfect and yet, it represents for one a symbol of the right part and a similar conviction seems to be the case for others. If one cannot dismiss the religious claim of the others nor assimilate it completely into one's own, a plausible alternative is to assume that all are different creeds, which in spite of meandering and crossing, actually run parallel to meet only in the ultimate at the very end of the human Pilgrimage. Religion will then be parallel parts, and one most urgent duty would be to develop one's own respective tradition so that everyone can meet at the end and in the depth of all traditions, in other world, one can be a better Christian, a better Marxist, and a better Hindus.

This attitude represents very positive advantages. It is tolerant; it respects others and does not judge them. It keeps the boundary clear, and spurs constant reform of one's own way. On the other hand, it is not free of difficulties. First of all, it seems to go against historical experience that the different religious and human tradition of the world has usually emerged from mutual interferences, influences and fertilization. It too hastily assumes that every human tradition has in itself, all the element for further growth and development. In one word, it assures the self-sufficiency of every tradition and seems to deny the need for mutual learning. Panikkar is hereby advocating an intra-religious dialogue as the model of authentic interpretation of any contemporary religious experience. In his book Intra-religious dialogue, he affirms the following about his personal journey:

> I left as a Christian, and I find myself a Hindus, and I return a Buddhist, without having ever ceases to be a Christian.

Panikkar's approach is fundamentally ecumenical. The encounter of world religion is constitutive of Christian theological reflection. Panikkar has recently underlined an emerging tendency in the comparative studies of religion which complementarily must be achieved within the individual himself.

Panikkar further affirms that:

> The burden of our tale is this: Any inter- religious and inter-human dialogues, any exchange among culture, has to be proceeded by an intra- religious intra- human dialogue an internal conversion within the person. We can only bridge the gap between east and west in this case, if we realize the synthesis and the harmony within that microcosm of ourselves.

Raimundo Panikkar's suggested three attitudes to the study of religion. They are Exclusivism, Inclusivism and Parallelism.

Mircea Eliade: The sacredness of nature and cosmic religion

Mircea Eliade, in his *The sacred and the profane*, contemplated a great deal on nature in the context of its sacredness. To him, nature is never only nature;

it is always fraught (filled) with religious values. He was emphatic that the cosmos is a divine creation; coming from the hands of gods. The world, to Eliade, is impregnated with sacredness. The opinion that the creation of the world has to do with divine connotation is also African only that Africans believe that the creation of the world is entirely the business of a Supreme Deity who has entrusted such responsibility into the hands of certain divinities which are subject to Him. The cosmos, the world of nature is not simply engendered by the gods, the gods did more; they manifested the different modalities of the world and of cosmic phenomena.

The world is seen in such a manner that by experience, and by observation, man, in his religiosity, discovers the many modalities of the sacred, and hence, of being. By modalities, we mean natural phenomena like the sun. To the Igbo of south east Nigeria, sun is not just a mere creature of God to give sunrays to mankind. They see beyond that. It has the essence of divinity. They call it Anyanwu, the sun divinity. The sun divinity is so important to the Igbo, to the extent that its role is often confused or identified with the activities of Chukwu, the Supreme Deity.

In the same manner, the Yoruba of south west Nigeria sees in oceans and rivers something more than natural phenomenon. The river is a modality impregnated with values in the Yoruba religious thought. Hence, they believe that Yemoja/ Olokun are the dual divinities in charge of the rivers. To them, these divinities are embodiments of wealth and riches. In the words of Eliade, the world exists with its structure, not a mere chaos. It is under the control of the Supreme Being and other divinities. The world presents itself as creation, as the work of God. It reveals itself in many aspects of the sacred. The sky, directly "naturally", reveals the infinite distance, the transcendence of the Deity. The earth too presents itself as universal mother and nurse. The cosmic rhythms manifest order, harmony and so on. The cosmic as a whole is living and sacred and reveals the modalities of being and sacredness.

In the words of Eliade, the sacred is revealed through the very structure of the world. To a religious man therefore, the supernatural is indissolubly connected with the natural that nature always expresses something that transcends it. This position expressed by Eliade is very much in agreement with Africans' beliefs. Consequently, to an African, a sacred stone is venerated because it is sacred, not because it is a stone; it is the sacred dimension manifested through the mode of being of the stone that reveals its true essence. In the African context, the sky shows itself to be infinite, transcendent. It is pre-eminently "wholly other" than man and his environment. Transcendence, to an African, is revealed by the fact that the sky is higher than them. "Most high" spontaneously becomes an attribute of divinity.

In the transcendence, what Eliade calls the "sidereal zones", dwells the God and the gods; there, a few privileged mortals make their way, by rites of ascent. The transcendental category of the height, of the super- terrestrial, of the infinite is revealed to the whole man, to his intelligence and his soul. That is why there is a vacuum in man that could only be filled by God. That is why when man beholds the sky, he simultaneously discovers the divine incommensurability and his own situation in the cosmos. For the sky, by its own mode of being, reveals transcendence force, eternity.

Eliade writes that celestially structured Supreme Beings tend to disappear from the practice of religion and from cults. They departed from among men; they withdrew to the sky and become remote, inactive gods (dei otiose). How correct is this opinion of Eliade in the context of African religion? Eliade has not expressed African position here. His opinion here is illogical when it comes to the interpretation of Africans belief or expression of Supreme Being. As Africans believe in the transcendental nature of God, they also believe in His immanence. That is, He is close to them. He could feel with them and he is sensitive to their plight here on earth. J.O. Awolalu and P.A. Dopamu, in their *West African Traditional Religion*, write that with reference to control and maintenance of the world, Africans think that, first, God is the creator. Second, God still creates and maintains the world. He is ever more interested in what goes on in the world of men. Even, the names of God among Africans such as Ngewo (Mende), Nyame (Akan) Osanobwa (Edo) show that God really maintains the universe.

God does not only maintain the universe; He also controls it. The sun, the moon, the stars, the created order in general and all that go on in the world are under His control. All these and the spirits, divinities and ancestors are always there and at all times, for His use. He can manifest Himself through any of these. The deities' authorities emanated from Him and are permitted to function by Him. It is also the belief of Africans that the supreme God is not only in control of all, but He is the absolute controller of the universe. In this perspective, the Ashanti of Ghana believe that Nyame is the Supreme Being upon whom men lean and do not fall.

The Yoruba of Nigeria's myth, recorded by Awolalu and Dopamu on God's control of the universe is an interesting one. According to the myth, 1700 divinities that were in the custody of the world in trust for God (*Olodumare*) conspired against Him, and demanded from Him the right to rule the universe on their own without reference to *Olodumare* for sixteen years. *Olodumare* advise them to first experiment for sixteen days and they agreed to this. They went away from the presence of the "wholly other" with pride and joy. As soon as they left, the Almighty switched off the machinery of the universe and everything ran contrary to its normal course. The divinities were about their new assignment, but since everything now ran against nature, they found the world unmanageable. The divinities failed within eight days of their secession. They realized their folly and their helplessness, and they summoned up courage and with drooped heads returned in penitence to the king of superlative attributes, and asked for His forgiveness.

Eliade also wrote in his book, *The sacred and the profane* that because of the remoteness of God, "Africans rarely made offering to him except during storm". It should be noted that Eliade has also expressed here, a narrow perspective of Africans religious thought. Definitely Africans hardly have cults for the Almighty God. There is no image of Him. Yet, they worship Him. They venerate Him. Sacrifice is also being offered to Him. Whenever an African man offers a sacrifice to his deity, he believes that a portion of the sacrifice would be reserved by that particular deity to the Supreme Being. Thus, indirectly, in this regard, the Africans offer sacrifice to the Supreme Being. The Urhobo of Nigeria could be sited as example of African people that offer direct veneration to the Almighty. Every morning, the head of the family or lineage takes a little quantity of "orhe" (white chalk) in his hand.

He breaks it down into powder with his thumb. While breaking the orhe into powder, he addresses Oghene (the Supreme Being) in His generic names and expresses his heart's desire to Him looking up steadily and intently into the horizon. He then blows the powdered white chalk into the air with a release of a deep breath which he had earlier taken.

Is Eliade correct when he writes that celestially structured supreme beings tend to disappear from the practice of religion and from cults of the Africans? Eliade has not expressed African position here. His opinion here is illogical when it comes to the interpretation of Africans belief or expression of Supreme Being. As Africans believe in the transcendental nature of God, they also believe in His immanence. That is, He is close to them. He could feel with them and he is sensitive to their plight here on earth. Their focus is on the Supreme Being even when they venerate the deities.

Bibliography

Agency for Cultural Affairs. (1972).'Japanese Religion: A Survey'. Tokyo: New York, and San Francisco kodansha International Ltd.

Anderson, N. Ed. (1975). *The world religions* Ground Rapids, MI: Win. B. Eardmana Publishing Co.

Babalola, S. F. (2019). *Inter-faith relation in Nigeria.* Akure: Akin-Olu Ventures.

Bidmos M.A. (1993). *Inter-religious dialogue: The Nigerian experience.* Isolo: Iroko printers.

Boa, K. (1977). *Cults, world religions and you.* Wheaton 11: Victor Books.

Burtt, E. A. Ed.(1955), *The teaching of the compassionate Buddha*, New York: New American Library.

Eliade, M. (1957). *The sacred and the profane.* New York: A Harvest HBJ Book.

Halt, P.M. (1970). *The Cambridge history of Islam* London: Cambridge press.

Hume, R. E. (1969). *The World's living religions* New York: Charles Scribner sons.

Lateju, F. T. (2002). *Guidelines for Christian witnessing among Muslims.* Ibadan: Baptist Press Limited.

Offner, C. B. and Anderson N. (1976). *The World's religion.* Grand Rapids.

Smart, N. (1969). *The religious experience of mankind.* New York: Charles Scribner and sons.

Smart, Ninian (1969), *The religious experience of mankind*, New York: Charles Scribner's sons.

Smith, Huston (1965), *The religion of man*, New York: Herper and Row.

Umejesi, I. O. (2010), *Oriental religions*, a quest for liberation and wisdom, Ekpoma: Pon publishers Ltd.

Zegers, Frans J. L. (1999). Are Allah and God the same? J P C Inc.

Publisher: Eliva Press Global Ltd

Email: info@elivapress.com

Made in United States
North Haven, CT
04 November 2026

21772742R00054